16th Anglo-French Physical Acoustics Conference (AFPAC 2017)

Journal of Physics: Conference Series Volume 1017

Marseille, France
23 – 25 January 2017

ISBN: 978-1-5108-6291-3
ISSN: 1742-6588

Printed from e-media with permission by:

Curran Associates, Inc.
57 Morehouse Lane
Red Hook, NY 12571

Some format issues inherent in the e-media version may also appear in this print version.

Printed by Curran Associates, Inc. (2018)

For permission requests, please contact the Institute of Physics at the address below.

Institute of Physics
Dirac House, Temple Back
Bristol BS1 6BE UK

Phone: 44 1 17 929 7481
Fax: 44 1 17 920 0979

techtracking@iop.org

Additional copies of this publication are available from:

Curran Associates, Inc.
57 Morehouse Lane
Red Hook, NY 12571 USA
Phone: 845-758-0400
Fax: 845-758-2633
Email: curran@proceedings.com
Web: www.proceedings.com

Table of contents

Volume 1017

16th Anglo-French Physical Acoustics Conference (AFPAC)

23–25 January 2017, Marseille, France

Accepted papers received: 30 April 2018
Published online: 11 May 2018

Preface

Papers

IOP Conf. Series: Journal of Physics: Conf. Series **1017** (2018) 011001 doi:10.1088/1742-6596/1017/1/011001

16th Anglo-French Physical Acoustics Conference (AFPAC 2017)

FOREWORD/PREFACE

The 16th Anglo-French Physical Acoustics Conference (AFPAC) was held on 23-25 January 2017 at the Laboratoire de Mécanique et d'Acoustique (LMA), Marseille, France. The event was organised by Dr Cécile Baron and Dr Philippe Lasaygues (LMA), who ensured a successful unfolding of both the conference sessions at the laboratory, and the social events in the city centre. The involvement of other staff from the LMA guaranteed that everything went very well and smoothly. A total of 50 papers were presented at the conference, which featured over 60 delegates. We would like to extend our thanks to Cécile, Philippe and the LMA staff for the pleasant times we had.

The first day of AFPAC kicked off with an invited presentation by Dr Claudia Fritz (Institut Jean Le Rond d'Alembert, Paris, France) who provided a fascinating insight into the qualities of Stradivarius versus modern-built violins, with surprising results. A diverse range of physical acoustics topics were broached on this day, ranging from the experimental ultrasonic characterisation of cultural heritage reproductions, acoustic emissions as a tribochemical sensing technique, acoustic characterisation of microbubble clouds and defect characterisation using ultrasonic arrays. The first of two Conference Dinners was held at Le Malthazar, in the Vieux-Port area of Marseille. The second day featured an invited presentation by Prof Geoff Nash (University of Exeter, UK) on surface acoustic wave-graphene interactions. Other works presented on this day included liquid foams, soap films and acoustic dissipation processes, nonlinear elastic metamaterials, transcostal high-intensity focused ultrasound for the treatment of liver tumours, torsional guided wave EMAT for SHM, the acoustic signature of skin samples for cosmetic applications and a range of numerical methods for NDE. An invited talk was given by Dr Frédéric Padilla (Labtau, INSERM, Lyon, France), who discussed the applications of ultrasound for bone tissue engineering and regenerative medicine. The second Conference Dinner was held at Les Arcenaulx, also in the Vieux-Port area, and offered a selection of locally sourced foods and organic wines. During the Conference Dinner, the 2016 Bob Chivers prize (a prize awarded by IOP / Physical Acoustics Group to students engaged in a PhD in a UK university) was awarded to Dr Anton Van Pamel (Imperial College London, UK) for his paper entitled "Finite element modelling of elastic wave scattering within a polycrystalline material in two and three dimensions".

On the final day of the conference, Prof Jeff Bamber (Institute of Cancer Research, London, UK) gave an invited talk reviewing elastography as an imaging modality for characterising the biomedical properties of tissue, in research and in clinical medicine. The conference concluded with a tour of the state-of-the-art facilities at the Laboratoire de Mécanique et d'Acoustique.

Dr Pierre Gélat (UCL, United Kingdom), IOP / Physical Acoustics Group, Chair,
Dr Alain Lhémery (CEA, France), SFA / GAPSUS,
Pr Nader Saffari (UCL, United Kingdom), IOP / Physical Acoustics Group.

IOP Conf. Series: Journal of Physics: Conf. Series **1017** (2018) 011002 doi:10.1088/1742-6596/1017/1/011002

Peer review statement

All papers published in this volume of *Journal of Physics: Conference Series* have been peer reviewed through processes administered by the proceedings Editors. Reviews were conducted by expert referees to the professional and scientific standards expected of a proceedings journal published by IOP Publishing.

IOP Conf. Series: Journal of Physics: Conf. Series **1017** (2018) 012001 doi:10.1088/1742-6596/1017/1/012001

Ultrasonic wave propagation in powders

R S Al-Lashi[1,2], M J W Povey[1], and N J Watson[3]

[1]School of Food Science and Nutrition, Faculty of Mathematics and Physical Sciences, University of Leeds, UK.
[2]Department of Engineering, School of Science and Technology, Nottingham Trent University, Clifton Lane, Nottingham, NG11 8NS, UK
[3]Department of Chemical and Environmental Engineering, University of Nottingham, UK.
Correspondence Email: raied.allashi@ntu.ac.uk

Abstract. Powder clumps (cakes) has a significant effect on the flowability and stability of powders. Powder caking is mainly caused by moisture migration due to wetting and environmental (temperature and humidity) changes. The process of moisture migration caking involves creating liquid bridges between the particles during condensation which subsequently harden to form solid bridges. Therefore, an effective and reliable technique is required to quantitatively and non-invasively monitor caking kinetics and effective stiffness. This paper describes two ultrasonic instruments (ultrasonic velocity pulse and airborne ultrasound systems) that have been used to monitor the caking phenomenon. Also, it discusses the relationship between the ultrasonic velocity and attenuation measurements and tracking caking kinetics and the effective stiffness of powders.

1. Introduction

Powders are the final or intermediate products of many food, chemical and pharmaceutical industries. The processing of powders can pose many challenges. Powder strength can increase significantly in the process equipment and during storage in the pack. This phenomenon is known as caking whereby free flowing powder is transformed into clumps or aggregates [1, 2]. It is mainly caused by moisture migration due to wetting and environmental (temperature and humidity) changes. The moisture migration caking process can be summarized in five steps [2]: (1) wetting and moisture absorption, (2) liquid bridging, (3) drying and moisture desorption, (4) hardening and solid bridging, and (5) compacting and caking. Powder caking can have detrimental effects on flowability, stability, solubility, mixing and dispersions [1]. As a result, it causes loss of product, delays in production, consumer complaints, storage and handling problems. Therefore, in situ monitoring of powder caking is required in many industries to improve their productivity and marketability. The ideal technique should be non-invasive, rapid, robust and effectively related to the caking stiffness and kinetics.

Ultrasonic measurements are easy to make and can be related to the elastic moduli of the bulk solids [3]. Several researchers used ultrasonic techniques to sense different properties of the powder in the past. For instance, Makse et al [4] studied the applicability of elasticity theory to granular materials. They performed acoustic experiments, numerical simulations and theoretical analysis to understand the elastic behavior of a deforming granular assembly of soft glass spheres (beads) under compressional load. Their results show good agreement between simulation and experimental results for the behavior of elastic moduli of the system. However, there was a significant disagreement in a system with shear perturbations. In this case both shear moduli and collective relaxation dynamics that are related to structural disorder and non-affine motion of grains should be included to obtain correct treatments. Coghill and Giang [5] investigated the possibility of using ultrasonic velocity to measure the strength of powder particles formed by agglomeration for two industrial materials: iron ore pellets, and calcined alumina. They demonstrated that ultrasonic velocity can be related to the packing geometry of coarser powder particles. Nevertheless, the velocity measurements are size independent and not repeatable for particle diameters less than 200 μm. Castellanos et al [6] studied the packing behavior of cohesive powder. They showed that the transmission of ultrasonic signal is determined by the elastic deformation at high consolidation stresses. Akseli and Cetinkaya [7] demonstrated air coupled (non-contact) ultrasonic technique for tablet characterisation. The coating thickness and mechanical properties of drug tablets were determined by this non-contact ultrasonic technique. Saggin and Coupland [8] used an ultrasonic device to monitor the dissolution kinetics of sucrose and lactose powder in water. They found that the reflected signal is sensitive to the amount of dissolved powder in water.

The aim of this paper is to investigate the feasibility of using ultrasonic techniques to monitor powder caking. The theory of ultrasonic wave propagation in powder will be discussed. Two ultrasonic instruments will be described. Experimental results for monitoring caking stiffness and kinetics will be presented that illustrates the relationship between sound velocity and powder caking.

2. Theory of ultrasonic wave propagation in powder

The acoustic properties of granular materials under confining pressure can be extremely nonlinear compared to continuum elastic solids. The simplest theories that describe sound propagation in powder confined by an external stress are the effective medium theory [4, 5]. In principle, these theories apply to uniformly sized spheres where the wavelength is long enough to ignore the effect of scattering from the powder particles. The ratio of sound wavelength to the particle diameter in the wavelength limit is usually greater than 10. The effective medium theories are based on the Hertz-Mindlin contact theory which assumes that the elastic moduli of the medium can be derived from the average particle to particle contact. These theories can be used to calculate the bulk and shear moduli as uniform pressured is applied to the powder particles. The longitudinal (P-wave) sound velocity and the elastic moduli of the isotropic materials are related by [4, 5]

$$V = \left(\frac{K+4/3\mu}{\rho_{dry}}\right)^{1/2} \tag{1}$$

where V is the longitudinal sound velocity (m/s), K is the bulk modulus (N/m^2), μ is the shear modulus (N/m^2) and ρ_{dry} is the density of dry particles (Kg/m^3). Makse et al. [4] derived a formula for the sound velocity in an assembly of glass spheres under stress by substituting the values of elastic moduli into equation 1 as follows

$$V = Z^{1/3}\phi^{-1/6}\left(\frac{6\pi P}{K_n}\right)^{1/6}\sqrt{\frac{3k_n}{20\pi\rho_{dry}}} \tag{2}$$

where Z is the average number of contacts per particle, ø is the volume fraction of the sample, and $k_n=4\mu/(1-v)$ is the effective stiffness (N/m^2) which is defined in terms of shear modulus (μ) of the solid particles and the Poisson ratio (v) of the solid particle material. Equation 2 shows that the effective stiffness of the caked powder can be calculated from measurements of sound velocity by assuming Z and ø do not significantly vary and determining ρ_{dry} from the received signal amplitude. The sound velocity in caked powder is higher compared to free flowing powder. This is because of the tortuous path of the propagated ultrasonic signal in the free flowing powder as shown in figure 1.

2.1. Relationship between dry and wet density

The reflection and transmission coefficients at the interface between air and powder surface can be calculated as follow

$$R = \frac{Z_{powder}-Z_{air}}{Z_{powder}+Z_{air}} \tag{3}$$

$$T = \frac{2Z_{powder}}{Z_{powder}+Z_{air}} \tag{4}$$

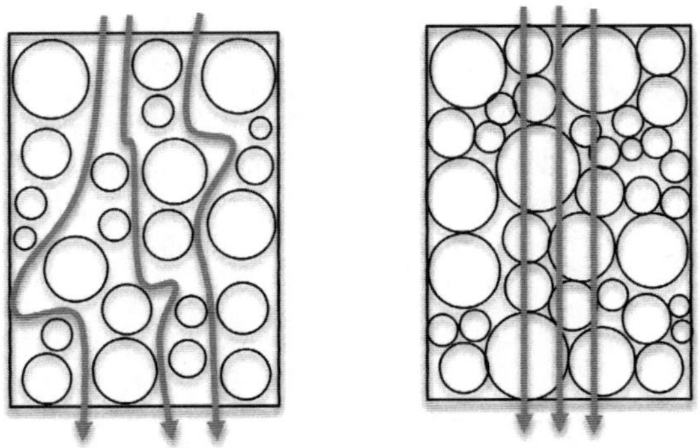

(a) Free flowing powder (b) Caked powder

Figure 1. Ultrasonic wave propagation in free flowing (a) and caked (b) powder. In comparison with caked powder (b), the ultrasonic signal takes longer to propagate in the free flowing powder (a) due to air gaps between particles.

where R and T are the reflection and transmission coefficients, respectively. The acoustic impedance of powder Z_{powder} and air Z_{air} are given by

$$Z_{powder} = \rho_{powder}\, V_{powder} \tag{5}$$

The approximate values of the density and sound velocity in air are 1.225 Kg/m^3 and 340 m/s, respectively. Thus, the acoustic impedance of air is

$$Z_{air} = \rho_{air}\, V_{air} \approx 417\ Kg/m^2 s \tag{6}$$

The sound velocity is approximately inversely proportional to the powder density [5] as follows

$$V_{powder} \approx \frac{1}{\sqrt{\rho_{powder}}} \tag{7}$$

Thus the powder acoustic impedance in equation 5 is equal to $\sqrt{\rho_{powder}}$. Substituting equation 5 into equations 3 and 4 yields

$$R = \frac{\sqrt{\rho_{powder}}-417}{\sqrt{\rho_{powder}}+417} \tag{8}$$

$$T = \frac{2\,(\sqrt{\rho_{powder}}\,)}{\sqrt{\rho_{powder}}+417} \tag{9}$$

Equations 8 and 9 can be used to calculate the reflection and transmission coefficient of dry and wet powder as follows

$$R_{dry} = \frac{\sqrt{\rho_{dry}}-417}{\sqrt{\rho_{dry}}+417} \tag{10}$$

$$T_{dry} = \frac{2\,(\sqrt{\rho_{dry}}\,)}{\sqrt{\rho_{dry}}+417} \tag{11}$$

IOP Conf. Series: Journal of Physics: Conf. Series **1017** (2018) 012001 doi:10.1088/1742-6596/1017/1/012001

$$R_{wet} = \frac{\sqrt{\rho_{wet}}-417}{\sqrt{\rho_{wet}}+417} \tag{12}$$

$$T_{wet} = \frac{2\,(\sqrt{\rho_{wet}})}{\sqrt{\rho_{wet}}+417} \tag{13}$$

The powder particles agglomerate as the powder is wetted. A larger portion of ultrasonic signal is reflected from the wet powder due to formation of liquid bridges which increase the reflection coefficient relative to air. Thus, comparing the peak to peak amplitude ($V_{p\text{-}p}$) and the reflection and transmission coefficients of the wet and dry powder yields

$$\frac{V_{p-p\,dry}}{V_{P-P\,wet}} = \frac{R_{dry}}{R_{wet}} \tag{14}$$

$$\frac{V_{p-p\,dry}}{V_{P-P\,wet}} = \frac{T_{dry}}{T_{wet}} \tag{15}$$

The relationship between wet and dry powder density can be obtained by comparing equations 14 and 15 with equations 10 to 13 and using Taylor series approximation as follows,

$$\rho_{wet} = \left(\frac{V_{p-p\,dry}}{V_{p-p\,wet}}\right)^2 \rho_{dry} \tag{16}$$

3. Ultrasonic measurement instruments

Two different instruments were used to perform the caking effective stiffness and kinetics measurements. The instruments are ultrasonic velocity pulse and airborne ultrasound systems. Both systems are operated in through-transmission (pitch-catch) mode. This approach is based on transmitting ultrasonic compression wave pulses between two piezoelectric transducers. However, each system has different setup and transducers as illustrated in sections 3.1 and 3.2.

3.1. Ultrasonic velocity pulse system

This system was used to perform the caking effective stiffness measurements. It consists of off-the-shelf instrument (Pundit Lab, Proceq Group, Switzerland) connected to a pair of transducers. The transducer pairs are placed in contact with the powder sample as shown in figure 2. Two different pairs of transducers were used in these measurements. Their centre frequencies are 24 and 50 KHz. five different thickness of powder samples were used to separate the transducer as shown table 1. A Lecroy oscilloscope (WaveRunner 44Xi, Teledyne LeCroy SA, Geneva - Switzerland) was connected to the instrument to monitor the received signals from the receiver transducer.

The system was calibrated by aligning the transducers and measuring the speed of sound in air. The powder sample was placed between the transducer pairs and wetted with water. Then it was left for some times to dry. The speed of sound and attenuation were measured during the drying time. This procedure was repeated for each sample in table 1.

3.2. Airborne ultrasound system

The interaction between the coating liquid and powder was monitored by using a bespoke airborne ultrasound system. This system is based on sending and receiving ultrasonic signal through air by a pair of non-contact transducer as shown in figure 3. The transducer centre frequency was 140 KHz and the path length between the transducers was 76 mm. This value was selected to be identical to the transducer focal distance. A high power amplifier was used to drive the transmitter transducer to transmit the signal through air and an automatic gain compensation amplifier was used to amplify the received signal from the receiver transducer. The transmitted and received signals were monitored via a handy scope HS3

USB oscilloscope (TiePie Engineering, Sneek, Netherlands) connected to the transducers. The sample was scanned automatically by connecting the transducers to a movable stage (Prior Scientific Instruments Ltd, Cambridge, UK). A high precision motor was embedded in this stage and controlled by ProScan III controller. LABVIEW software (National Instruments Corporation (U.K.) Ltd., Newbury, UK) was used for automation control.

The powder sample shown in figure 3 was non-ionic surfactant linear alkylbenzene sulfonates (LAS) flake powder which is one of major chemicals used in manufacturing detergent powder. The powder was tableted to obtain a single surface particle to ignore the effects of changing particle shapes and sizes on studying particle-liquid interaction. The calibration was made by sending and receiving the signal from the stage without the tablet. This signal is used as a reference to compare with the waveform obtained from placing the tablet between the transducers as shown in figure 4. It can be seen that the signal amplitude of the reflected signal from the tablet surface has been decreased significantly. In addition, the time of flight of the reflected signal is less than the reference time. This is because part of ultrasonic signal is reflected from the top surface to form the first echo. The other portion of this signal penetrates into tablet and reflects from its bottom surface to form the second echo.

The reduction in the signal amplitude triggers the gain compensation amplifier to increase gain to amplify the received signal. This indicates that the peak to peak signal amplitude is inversely proportional to the instrument gain. Thus, using dry and wet powder gain (G_{dry} , G_{wet}) in equation 16 instead of the corresponding amplitude ($V_{p\text{-}p\ dry}$,$V_{p\text{-}p\ wet}$) yields

$$\rho_{wet} = \left(\frac{G_{wet}}{G_{dry}}\right)^2 \rho_{dry} \tag{17}$$

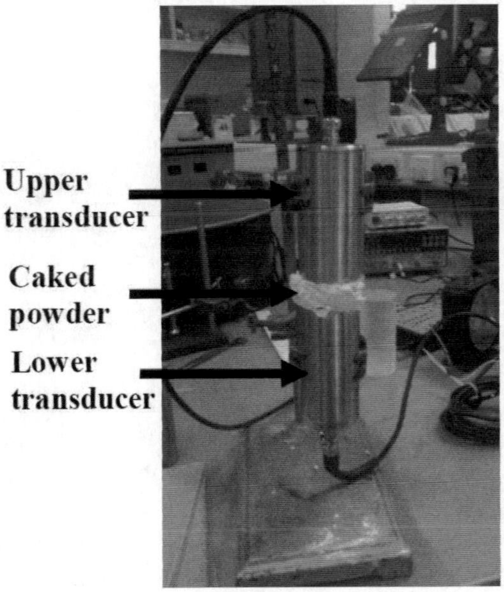

Figure 2. Ultrasonic transducers in contact with caked powder to measure sound velocity and attenuation.

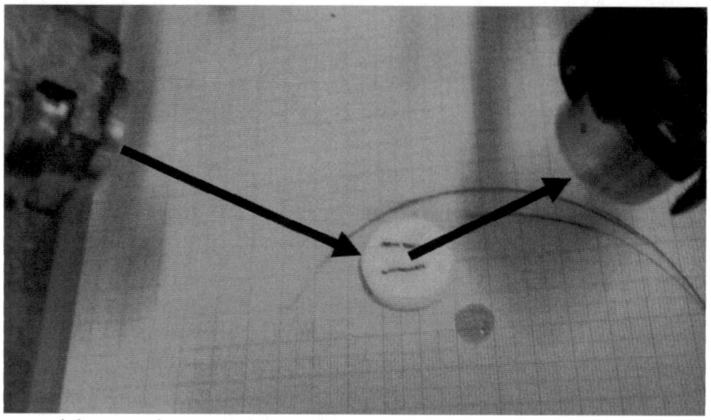

Figure 3. Non-contact airborne ultrasound transducers transmitting and receiving ultrasonic signal from the surface of LAS flake powder surface

Table 1. Different combination of ultrasonic transducers and powder thickness used in the Ultrasonic velocity pulse system measurements shown in figure 2.

Sample	Transducer center frequency (kHz)	Powder thickness (mm)
1	24	16
2	24	21.6
3	24	20
4	50	19
5	50	24

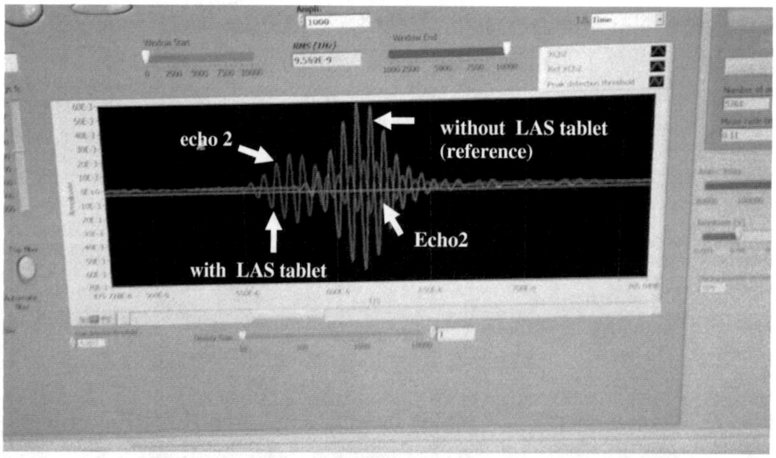

Figure 4. Comparison of the reflected signals that received by the airborne ultrasound transducers shown in figure 3. Echo 1 (red color) corresponds to the signal reflected from the stage surface without placing LAS flake tablet. However, echo 2 (violet color) represents the reflected signal from the LAS tablet.

4. Results and discussion

4.1. Cakes stiffness monitoring

The particles are agglomerated as the powder was wetted due to formation of liquid bridges between these particles. The speed of sound in the wetted powder is low as shown in figure 5. This is because these liquid bridges do not readily transmit sound due to their very small shear modulus and hence acoustic mismatch with the solid particles (High shear modulus) which they connect. However, solid bridges are formed between the particles as the sample consistently dries from 35ºC to 22ºC. Therefore, the speed of sound significantly increases by an order of magnitude as shown in figure 5. This increase in sound speed depends on the transducer centre frequency. The speed of sound is lower in the higher frequency transducers.

The peak to peak voltage was also measured with the drying time as shown in figure 6. The initial portion of the curves is similar to the velocity curves due to liquid bridges. As the sample dries, the peak to peak voltage of the signal increases due to formation of solid bridges. This indicates that there is a reduction in attenuation across the sample.

The minimum size of the detected caked regions is of the order of a wavelength (λ) of sound ($v=f\lambda$). The wavelength is 3 mm in the dry powder at 24 KHz frequency (f). However, it increases up to 175 mm in the wet caked powder. The transition from dry to caked powder occurs at a wavelength of around 9 mm.

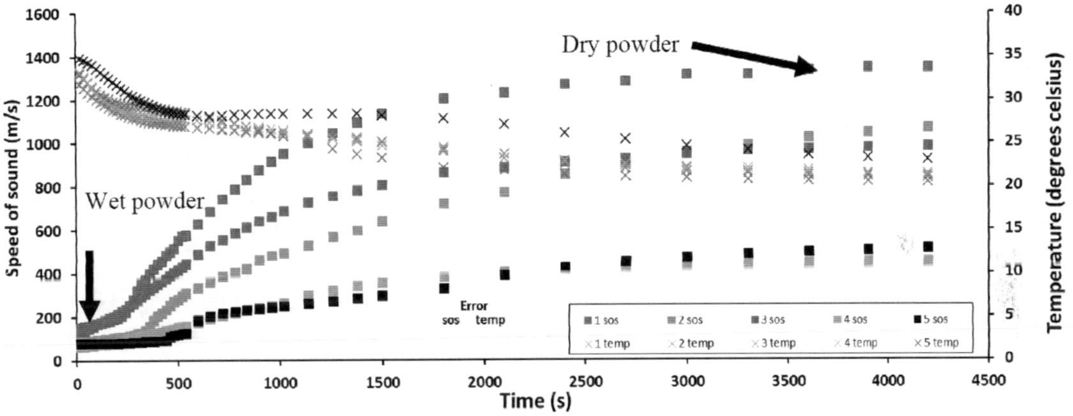

Figure 5. Ultrasonic velocity measurements obtained by the system shown in figure 2. The sound speed is low in wetted powder whilst it dramatically rises in the dried powder. The symbol sos corresponds to the speed of sound for the samples shown in table 1 and the symbol temp corresponds to the temperature.

IOP Conf. Series: Journal of Physics: Conf. Series **1017** (2018) 012001 doi:10.1088/1742-6596/1017/1/012001

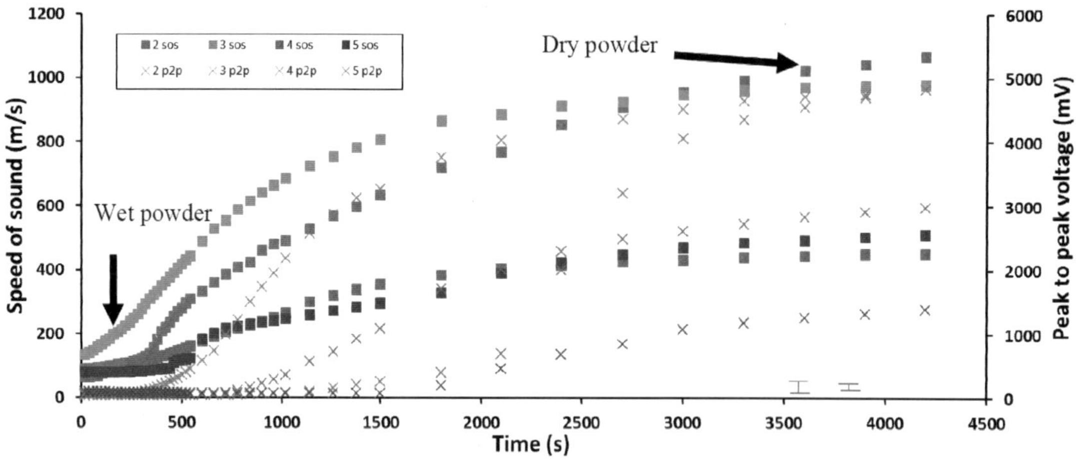

Figure 6. Ultrasonic attenuation measurements obtained by the system shown in figure 2. The attenuation is high in wetted powder whilst it greatly improves in the dried powder. The symbol sos corresponds to the speed of sound for the samples shown in table 1 and the symbol p2p corresponds to the peak to peak signal.

4.2. Cakes kinetics monitoring

The interaction between non-ionic surfactant (Neodol) with ionic surfactant was monitored by adding a droplet of Neodol to the surface of LAS tablet. The tablet surface was scanned to measure the pulse time of flight and signal amplitude. Figure 7 shows 3D image of the powder surface before and after adding Neodol droplet. The gain is high before adding the droplet. This is because the signal is highly attenuated as it penetrates through the tablet. This triggers the gain compensation amplifier of the instrument to increase the gain automatically to compensate this reduction in signal amplitude. However, liquid bridges are formed between the particles after adding the liquid droplet. As a result, large portion of the signal is reflected from the surface and therefore the gain value is small. Nevertheless, the gain starts to increase again as the liquid droplet dries at the surface and ultrasonic signal penetrates through the sample.

IOP Conf. Series: Journal of Physics: Conf. Series **1017** (2018) 012001 doi:10.1088/1742-6596/1017/1/012001

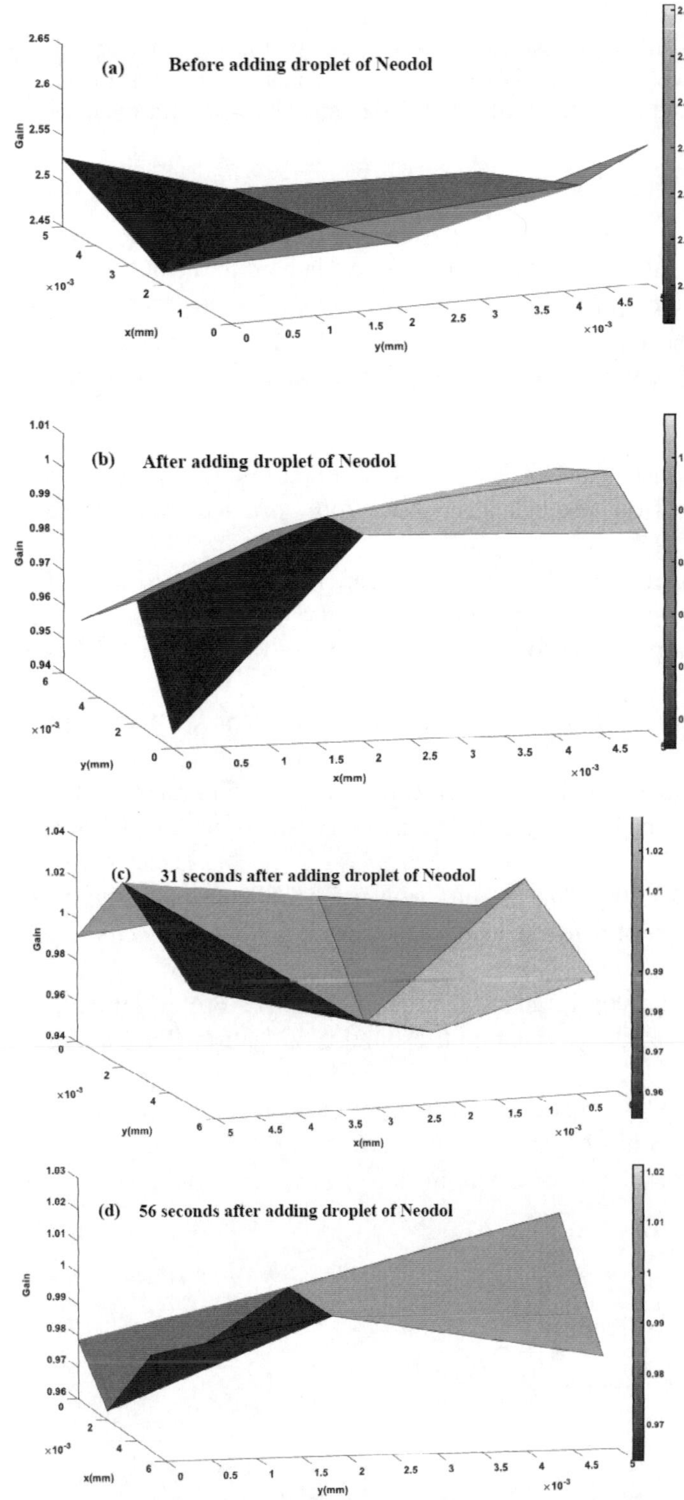

Figure 7. 3D images of LAS tablet obtained: (a) before adding Neodol droplet, (b) directly after adding Neodol droplet, (c) 31 seconds after adding Neodol droplet, and (d) 56 seconds after adding Neodol droplet.

5. Conclusions

The goal of this paper was to provide an initial assessment of the feasibility of using sound speed and attenuation measurements to monitor powder caking. The results show that sound speed and attenuation are sensitive to cake effective stiffness and the kinetics of caking. We conclude that ultrasound can be used to monitor powder caking.

Acknowledgement
This work was supported by AMSCI funding of Chariot Consortium, particularly the Advance Manufacturing Supply Chain Initiative which is a government supply chain fund for helping to rebuild British manufacturing process. Also the authors would like to EPSRC grant (grant number EP/M026310/1) for their support. The authors would like to thanks Mr. Hossam Tantawy and Ms. Claire Duckitt from Procter & Gamble, UK for supplying us with chemicals and useful discussions.

6. References

[1] Thakur, S. C., Ahmadian H., Sun J. and Ooi J. Y. 2014 An experimental and numerical study of packing, compression, and caking behaviour of detergent powders *Particuology* **12** 2–12.
[2] Christakis, N., Wang J., Patel M. K., Bradley M. S. A., Leaper M. C. and Cross M. 2006 Aggregation and caking processes of granular materials: continuum model and numerical simulation with application to sugar *Advanced Powder Technol.* **17** 543–565.
[3] Povey, M. J. W. 1997 *Ultrasonic techniques for fluids characterization*, Academic press San Diego, California, USA.
[4] Makse, H. A., Gland N., Johnson D. L. and Schwartz L. 2004 Granular packings: Nonlinear elasticity, sound propagation, and collective relaxation dynamics *Phys. Rev. E* **70**.
[5] Coghill, P. J. and Giang P. 2011 Ultrasonic velocity measurements in powders and their relationship to strength in particles formed by agglomeration *Powder Technol.* **208** 694–701.
[6] Castellanosa, A., Jiab X., Soria-Hoyoa C. and Valverdea J. M. 2011 Probing the nature of the contact between fine particles by using ultrasound propagation *Particuology* **9** 659–662.
[7] Akseli, I. and Cetinkaya C. 2008 Acoustic testing and characterization techniques for pharmaceutical solid dosage forms *J Pharm Innov* **3** 216-226.
[8] Saggin, R. and Coupland J. N. 2002 Ultrasonic Monitoring of Powder Dissolution *J. Food Sci.* **67** 1473–1477.

IOP Conf. Series: Journal of Physics: Conf. Series **1017** (2018) 012002 doi:10.1088/1742-6596/1017/1/012002

Acoustic identification of the elastic properties of porous and nonporous superconducting materials $DyBa_{2-X}Sr_XCu_3O_{7-\delta}$

N Sayoud[1], H Lazri[1], E Ogam[2], Y Boumaiza[1] and A Boudour[1]

1. Laboratoire LEAM, Département de Physique, Faculté des sciences, Université Badji-Mokhtar, Annaba, Algérie.
2. Aix Marseille Univ, CNRS, Centrale Marseille, LMA UPR7051, Marseille, France
E-mail : ogam@lma.cnrs-mrs.fr

Abstract. This work involves the development of a method using the principle of acoustic microscopy to determine the elastic properties of high-temperature superconducting materials, the method is applied to analyse the variation of the elastic properties of the superconducting alloy $DyBa_{2-x}Sr_xCu_3O_{7-\delta}$ for different variations of the concentration parameters ($x=0$, $x=0.3$ and $x=0.6$), porosity and temperature. The method is based on the reconstruction of the reflection coefficient calculated from the acoustic signature of the signal received by the microscope during the exploration of the superconducting material for different concentrations. This permitted the determination of the velocities of the surface and volume waves from the modelled reflection coefficient. On the other hand, the elastic parameters of the material such as Young's, shear and bulk moduli were also deduced.

1. Introduction

The goal of this study is to develop a method that can improve knowledge on the underlying mechanisms that control high-temperature superconductivity and how to evaluate the enhancement of this property when some parameter of the material is modified. The nature of high-temperature superconductivity mechanisms still remains an open question. In non-destructive testing, the elastic constants of solids have been investigated by measuring the velocity and attenuation of transmitted or/and reflected ultrasound waves. Since there is essentially no difference between thermal motion (random vibrations in solids especially metals) and elastic waves it is possible to get information on the order of the mean free path of the electrons in solids.

Therefore, by recovering the elastic constants we can gain a better understanding about the interactions between electrons and the atomic lattice, lattice instabilities and phase transitions (the contribution of the free electrons to the sound attenuation vanishes completely due to recombination of the free electrons into Cooper pairs). Among the multiple techniques used in Non-Destructive Testing (NDT), we have chosen to do the study using acoustic microscopy technique as proposed by Quate et al [1,2,3] and further developed by Briggs et al [3].

The acoustic microscopy measurement technique is based on the emission and reception of ultrasonic waves [4], it allows the determination of the different elastic parameters of the superconducting material, to understand the mechanism of superconductivity [5], using the reflection coefficient and acoustic signature for this type of material at ultrasonic frequencies.

2. Theory

This work is done at high frequency (600 MHz) and therefore to allow the transmission and consequently the propagation of the ultrasonic waves in the solid, a coupling fluid was used as propagation medium. This is because the high frequency ultrasonic acoustic waves are very quickly damped by air. Water was employed as the coupling fluid. The acoustic wave propagating in the coupling fluid in the direction of the solid is considered as incident wave. It arrives at the surface of the solid ($DyBa_{2-x}Sr_xCu_3O_{7-\delta}$) at an angle of incidence θ . Part of the incident wave is transmitted into the solid while the other is reflected towards the liquid. The boundary conditions at the solid-fluid interface allow us to determine the reflection coefficient which is given by the following expression [6, 7 and 8]:

$$R(\theta) = \frac{Z_L \cos^2 2\theta_S + Z_S \sin^2 2\theta_S - Z_0}{Z_L \cos^2 2\theta_S + Z_S \sin^2 2\theta_S + Z_0}, \tag{1}$$

where Z_0 is the acoustic impedance of the fluid, Z_L and Z_S and are respectively the longitudinal and transverse acoustic impedances of the solid. The impedances are given by,

$$Z_0 = \frac{\rho_f c_f}{\cos(\theta)}, \quad Z_L = \frac{\rho c_L}{\cos(\theta_L)}, \quad Z_L = \frac{\rho c_S}{\cos(\theta_S)},$$

where ρ_f is the fluid density and c_f is the wave velocity in the fluid, c_L is the longitudinal wave velocity, c_S is the transverse wave velocity, θ_L and θ_S are the longitudinal and transverse wave angles in the solid. The relationship between the angles are given by,

$$k_0 \sin(\theta) = k_L \sin(\theta_L) = k_S \sin(\theta_S),$$

where k_0 is the longitudinal wave number, k_L and k_S are the longitudinal and shear wave numbers respectively in the solid.

The expression for the reflection coefficient $R(\theta)$ is used to determine the acoustic signature $V(z)$ according to the following equation [8]:

$$V(z) = \int_0^{\theta_{max}} p^2(\theta) R(\theta) \exp(2jk_0 z \cos\theta) \sin\theta \cos\theta d\theta \tag{2}$$

where θ_{max} is the maximum aperture of the lens, $p^2(\theta)$ is the pupil function.

The acoustic signature V(z) is composed of periodic signals with a periodicity Δz of interferences that give information on the acoustic surface wave. This period depends on the velocity of the surface waves according to:

$$V_R = V_{liq}\left[1-\left(1-\frac{V_{liq}}{2f\Delta z}\right)^2\right]^{-1/2} \tag{3}$$

The acoustic signature is associated with the values of the elastic parameters characterizing the material such as Young's modulus (E), shear modulus (G), bulk modulus (B) according to Eqns. (4, 5 and 6). It is also related to the structure and nature of the material [7,9].

$$E = \rho V_S^2 \frac{3V_L^2 - 4V_S^2}{V_L^2 - V_S^2}$$
$$G = \rho V_S^2$$
$$B = \frac{E}{3(1-2\upsilon)} \qquad (4,5,6,7)$$
$$\upsilon = \frac{2V_S^2 - V_L^2}{2\left(V_S^2 - V_L^2\right)}$$

where ρ is the density of material, V_S is the transverse wave velocity, V_L is the longitudinal velocity and υ is the Poisson ratio.

For the porous material with a porosity rate p, wave velocities are modified. The change is computed by replacing in the reflection coefficient function the modified propagation wave velocities of the volume waves and the density of the porous material by [10]:

$$V_L = V_{L_0}\left(1-p\right)^m$$
$$V_S = V_{S_0}\left(1-p\right)^m \qquad (8,9,10)$$
$$\rho = \rho_0\left(1-p\right)$$

where m is an empirical constant and the subscript 0 denotes a material with zero porosity.

3. Results and Discussion

3.1. Study of the superconducting material $DyBa_{2-x}Sr_xCu_3O_{7-\delta}$ $\left(x=0, x=0.3, x=0.6\right)$:

Our study is based on the modeling of the reflection coefficient as a function of the angle of incidence according to Eqn. (1) and the acoustic signature V (z) according to Eqn. (2). A computer program was developed to compute the acoustic signature V (z) using the reflection coefficient as a function of the angle of incidence ($R(\theta)$) computed using the measured longitudinal and shear wave velocities (Eqn. 1). The velocities were measured for different concentration parameter values (x = 0, x = 0.3 et x = 0.6).

3.1.1. The reflection coefficient $R(\theta)$:

The values of the longitudinal and transverse wave velocities of the superconducting material $DyBa_{2-x}Sr_xCu_3O_{7-\delta}$ as a function of the temperature were determined from the curves of the changes in the reference velocities found in [11]. The reference velocities were first recalculated then the reflection coefficients R (θ) for different values of temperature (T) from 80 ° K to 230 ° K, were determined. An example of some reflection coefficient curves (variation of the amplitude and phase of $R(\theta)$) for this material at 80 ° K for different concentrations are shown in Figure (1).

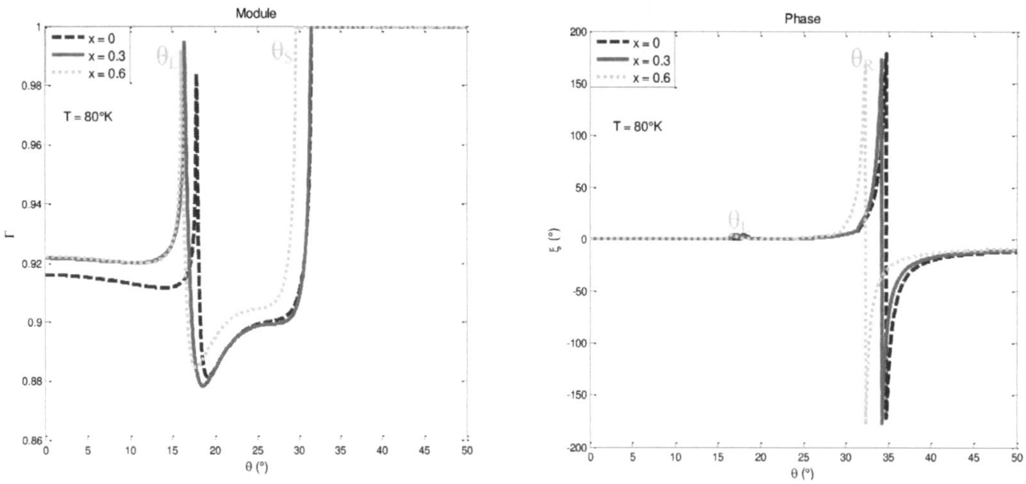

(a) (b)

Figure 1 : Variation of the amplitude and phase of $R(\theta)$ of the material $DyBa_{2-x}Sr_xCu_3O_{7-\delta}$ $(x=0, x=0.3, x=0.6)$ at 80°K - (a) Amplitude, (b) Phase.

It can be seen in Figure (1), that the amplitude (modulus) of the reflection coefficient varies from 0 to 1 and has two critical angles, corresponding to the angles θ_L (longitudinal critical angle) and θ_S (transverse critical angle). The phase curve contains two peaks, one very small corresponds to the longitudinal critical angle θ_L and the other, the largest, corresponds to the critical Rayleigh angle θ_R . For the different values of the parameter x, there is a significant shift of the critical angles to the lower angles for all modes with respect to the critical angles relative to the material which corresponds to the zero concentration value (x = 0).

3.1.2. The acoustic signature $V(z)$:

The determination of the reflection coefficient enables the determination of the acoustic signature $V(z)$. The acoustic signature curves for different concentration values at 80 ° K are depicted in Figure (2).

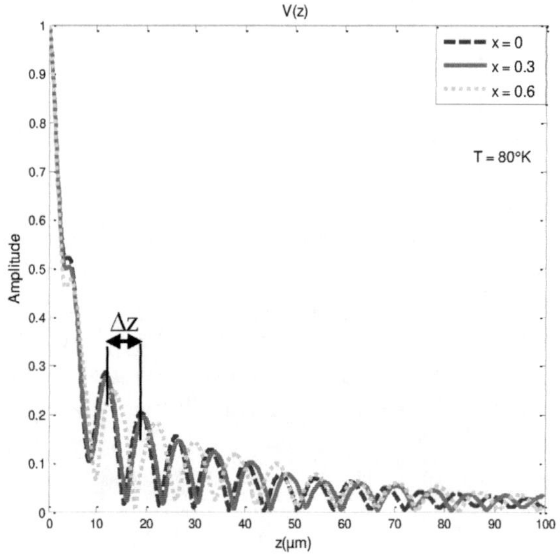

Figure 2 : The acoustic signature $V(z)$ of the material $DyBa_{2-x}Sr_xCu_3O_{7-\delta}$ $(x=0, x=0.3, x=0.6)$ at 80°K.

It will be noted that, depending on the parameter x, different curves of the acoustic signature are obtained with a decrease in the period of the pseudo-oscillations, which indicate a change in the velocities of the modes which propagate in the superconducting material studied.

The critical angles of the propagating modes at oblique angles of incidence takes into account: the longitudinal critical angle, the transverse critical angle and the velocity in the liquid $V_{Liquide}$. This makes it possible to obtain the velocities of these modes for the material studied using the Snell-Descartes law [8]:

$$V_L = \frac{V_{Liquide}}{\sin(\theta_L)}, \qquad V_S = \frac{V_{Liquide}}{\sin(\theta_S)}.$$

The following figures show the variation of the longitudinal and transverse velocities as a function of the temperature for different values of concentration x:

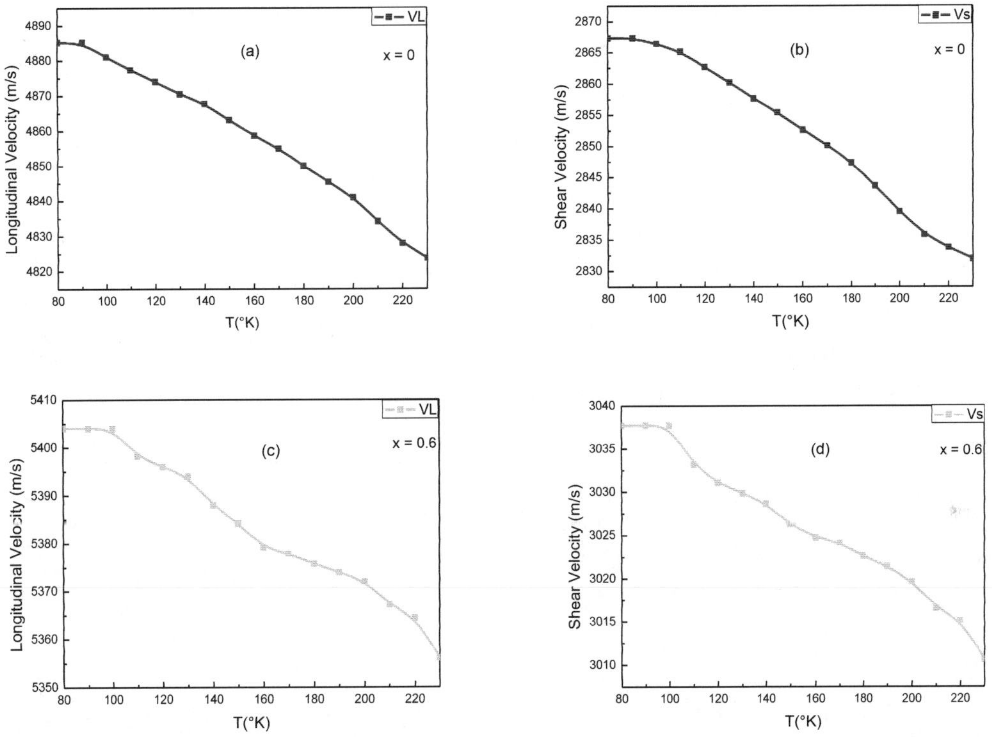

Figure 3: Variation of the longitudinal and transverse wave velocities (V_L and V_T respectively) as a function of temperature for the material $DyBa_{2-x}Sr_xCu_3O_{7-\delta}$ – (a) and (b) (x= 0), (c) and (d) (x=0.6) .

It is found that the velocities of the modes which propagate in the superconducting material study decrease with the increase of the temperature. The increase in the concentration x causes the increment of these velocities. These results compare very well with those in references (Yahya, 2008).

By introducing the values of the longitudinal and transverse velocities into the equations (4, 5 and 6), the elastic parameters (E, G, B) of the superconducting material considered as a function of the temperature can be determined for the three concentration values (x = 0, x = 0.3 et x = 0.6). The results obtained are presented in the following figures:

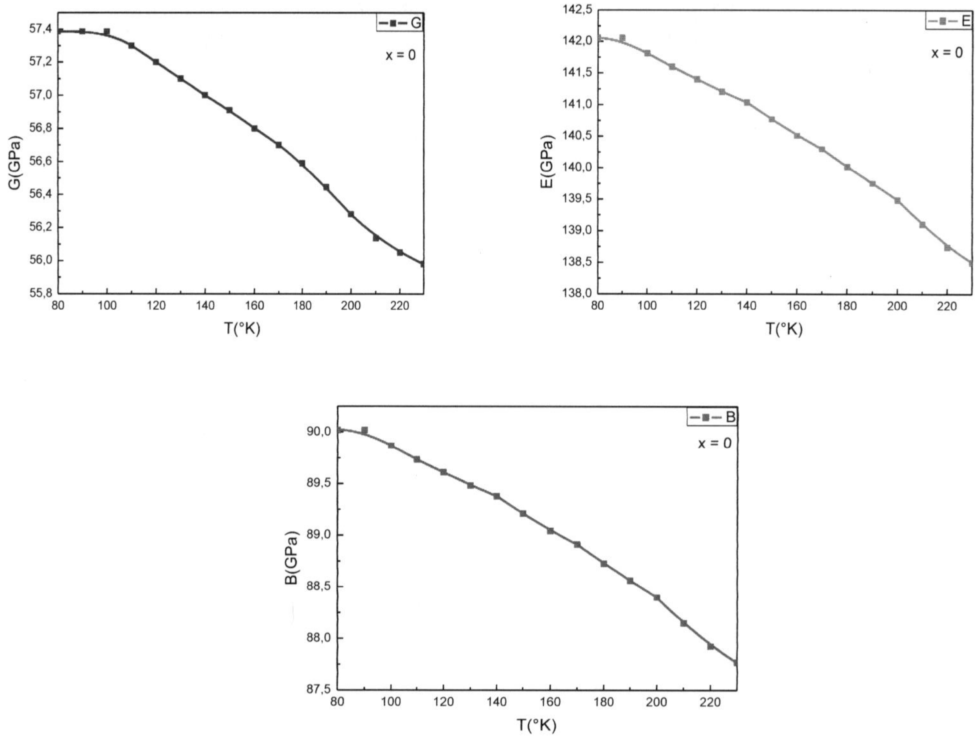

Figure 4: Variation of the elastic moduli (E, G, B) of $DyBa_{2-x}Sr_xCu_3O_{7-\delta}$ as a function of the temperature for the concentration x = 0.

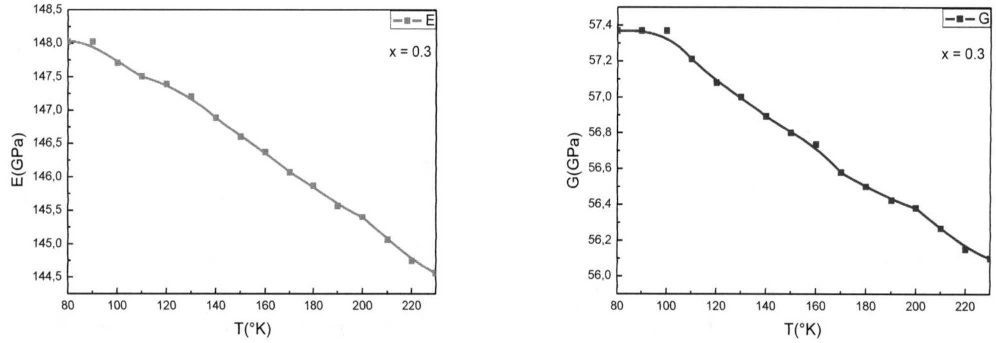

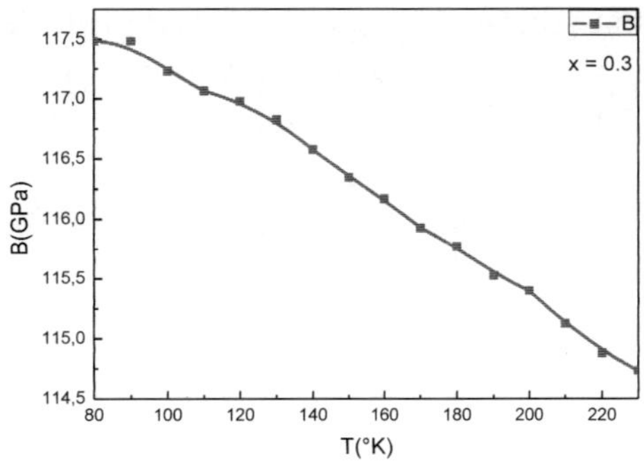

Figure 5: Variation of the elastic moduli (E, G, B) of $DyBa_{2-x}Sr_xCu_3O_{7-\delta}$ as a function of the temperature for the concentration x = 0.3

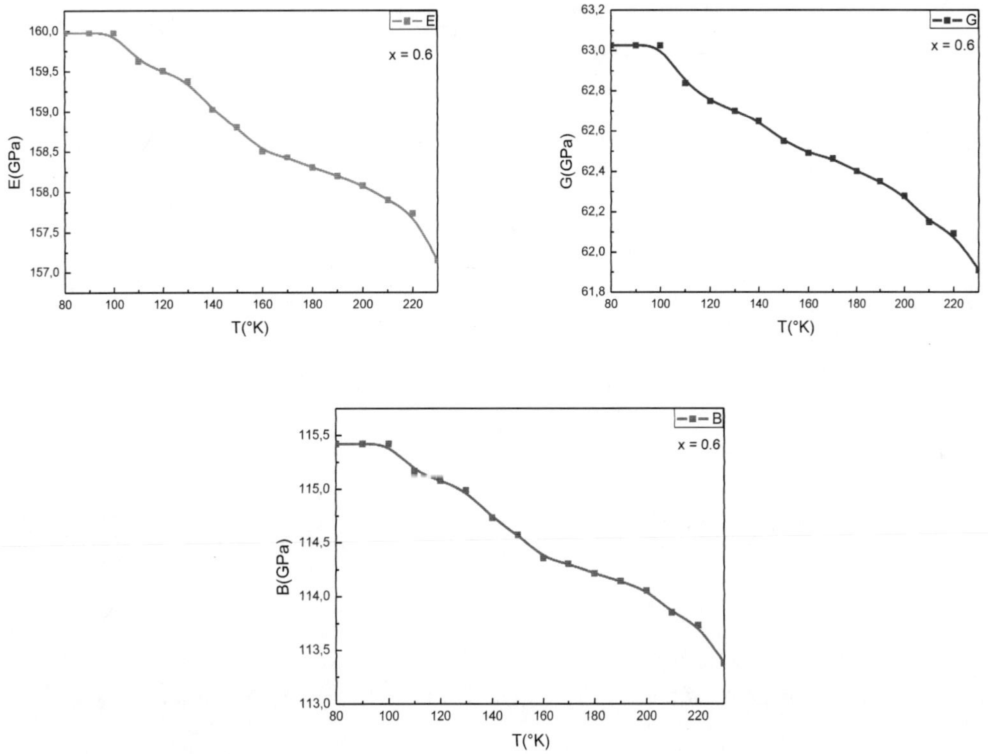

Figure 6: Variation of the elastic moduli (E, G, B) of $DyBa_{2-x}Sr_xCu_3O_{7-\delta}$ as a function of the temperature for the concentration x = 0.6

3.1.3. The reflection coefficient for porous $R(\theta)$ *:*

For the porous material the variations of the reflection coefficient were determined in the same way as previously with the replacement of the propagation velocities of the volume waves and the density of the porous material from the relations (7, 8 and 9). The variation of the reflection coefficient as a function of the porosity is shown in Figure (7).

IOP Conf. Series: Journal of Physics: Conf. Series **1017** (2018) 012002 doi:10.1088/1742-6596/1017/1/012002

There is a significant shift of the critical angles for all modes to the upper angles with respect to the critical angles of the non-porous material (p = 0%), which implies a decrease in the velocities of these modes. The reflection function gives the acoustic signature of the superconducting material for the parameter value (x = 0) and for different porosities (Figure 8).

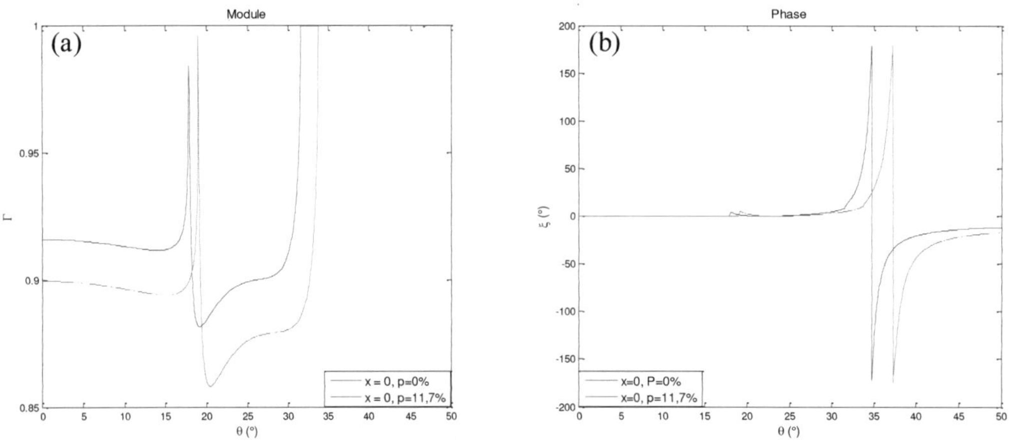

Figure 7: Variation of the reflection coefficient $R(\theta)$ of porous $DyBa_{2-x}Sr_xCu_3O_{7-\delta}$ for x=0 - (a) Amplitude, (b) Phase.

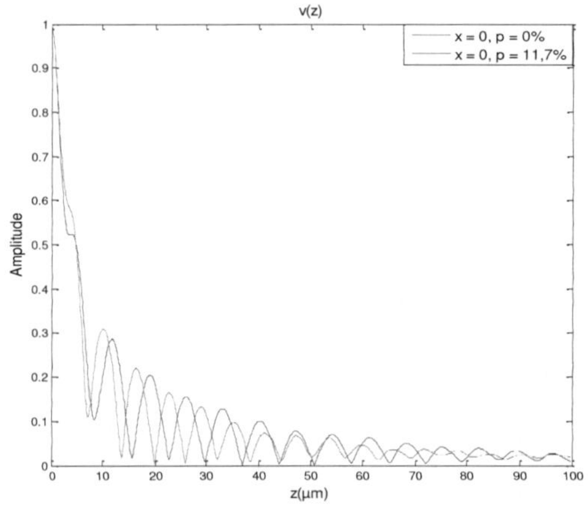

Figure 8 : Acoustic signature of porous $DyBa_{2-X}Sr_XCu_3O_{7-\delta}$ for p=0% and p=11.7% (x = 0).

From the acoustic signature V(z), the velocities of the modes propagating in the porous superconducting material studied and the elasticity moduli as a function of the porosity for the three parameters (x = 0, x = 0.3 and x = 0.6) are determined (Figure 9).

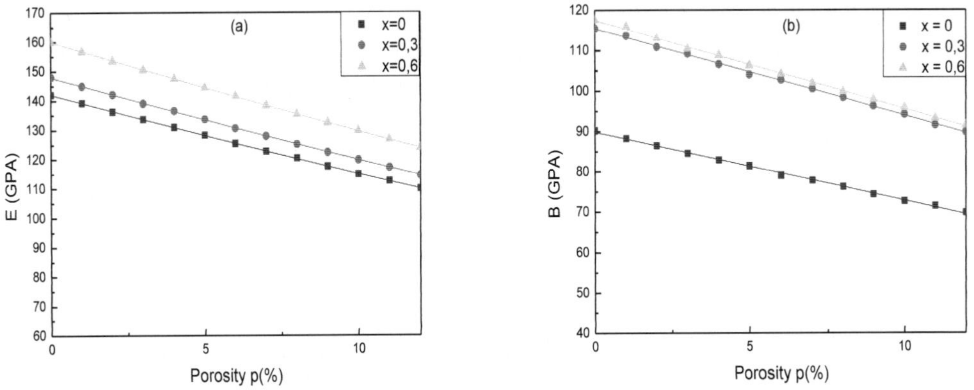

Figure 9: Variation of the elastic constants (E and B) of the porous $DyBa_{2-x}Sr_xCu_3O_{7-\delta}$ as function of porosity and the three concentration parameters (x = 0, x = 0.3 and x = 0.6) (a) E, (b) B.

Figure 9 shows that the elastic constants decrease with porosity for example, x = 0 and p = 0% Young's modulus E = 142, 05GPa, while for x = 0 and p = 11.7%, E = 110,91 GPa. The decrease in these parameters are due to the increase of porosity (Equations 8, 9) resulting in the dimunition of the longitudinal and transversal wave speeds and the reduction of the amplitude level of the received signal [10].

4. Conclusion

The change of concentration (parameter x) of the superconducting material $DyBa_{2-x}Sr_xCu_3O_{7-\delta}$ leads to a variation in the velocities of the modes propagating in this material, i.e. the increase of the concentration x leads to the increment of the longitudinal and transverse mode velocities. As a result, the moduli of elasticity change when the concentration number x of the superconducting material varies from 0 to 0.3 and then to 0.6.

For the porous superconducting material, the velocities are influenced by the increase of the porosity rate, which results in a decrease in the values of the elastic moduli of this material (Young's modulus, shear modulus and the bulk modulus. From the knowledge of these moduli, the longitudinal and and transverse wave velocities, the acoustic Debye temperature can be calculated [11]. The acoustic Debye temperature of a superconducting material can provide information about the role of phonons in its superconducting mechanism.

5. References

[1] C. F. Quate, A. Atalar, and H. K. Wickramasinghe. Acoustic microscopy with mechanical scanning-a review. Proc. IEEE **67**, 1092 (1979).

[2] R. A. Lemons and C. F. Quate. A scanning acoustic microscope. in IEEE Ultrasonics Symposium proceedings (1973), p. 18.

[3] X. D. Deng, T. Monnier, P. Guy and J. Courbon. Acoustic microscopy of functionally graded thermal sprayed coatings using stiffness matrix method and Stroh formalism. Journal of Applied Physics, 2013, 113(22), pp.224508:1-10.

[4] T. Tahraoui, S. Debboub, Y. Boumaiza, A. Boudour. Ultrasonic investigation for coated materials. Int. J. Microstructure and Materials Properties, Vol. 7 , Nos. 2/3 , 2012 .

[5] M.B. Solunke, K.B. Modi, V.K. Lakhani, K.B.Zankat, P.U. Sharma, P.V. Reddy et S.S.Sah. Effect of Ag^+addition on elastic behaviour of $Bi-2212$ superconductors. Indian Journal of Pure et Applied Physics. Vol. 45 , September 2007 , pp. 764 – 766 .

[6] L.M.Brekhovskikh, O.A.Godin. Acoustic of layered media I: plane and quasi-plane waves. Springer-Verlag, Berlin, (1990).

[7] A. Benbelghit, D. Boutassouna, B. Helifa, I. K Lefkaire. Determination of the elastic properties of some coted Materials by simulation of the analogue signal of the reflection acoustic microscope. NDT&E International 39(2006)76–81.

[8] G.A.D.Briggs. Acoustic Microscopy. Oxford University, 2010.

[9] P. Zinin and W. Weise, "Theory and applications of acoustic microscopy", in T. Kundu ed., Ultrasonic Nondestructive Evaluation: Engineering and Biological Material Characterization. CRC Press, Boca Raton, chapter 11, 654-724(2004).

[10] R. J. M. Da Fonseca, J. M Saurel, A. Faucaran, E.Massone, T. Taliercio, J. Camassel. Acoustic microscopy investigation of porous silicon. Elsevier science, thin solid films 255 (1995)155-158.

[11] N.A. Rasih, A.K. Yahya, Effet of Ba -site substitution by Sr on ultrasonic velocity and electron-phonon coupling constant of $DyBa_{2-x}Sr_xCu_3O_{7-\delta}$ superconductors, J. Alloys Compd. (2009), doit: 0.1016/J. jallcom. 02055, (2009).

The coherent shear wave in suspensions

M Alam[1], F Luppé[1], V J Pinfield[2] and P Maréchal[1]

[1] Laboratoire Ondes et Milieux complexes, Normandie Univ, UNILEHAVRE, CNRS, LOMC, 76600 Le Havre, France
[2] Chemical Engineering Department, Loughborough University, Loughborough, UK, LE11 3TU

francine.luppe@univ-lehavre.fr

Abstract. We consider a disordered suspension of spherical silica particles in water. For a particle size of a few hundred nanometres and concentration (volume fraction) around 0.15 to 0.2, experiments conducted in the MHz range have shown that the non ideal nature of water must be taken into account for the "longitudinal" coherent wave attenuation to be understood, because of wave conversions, from longitudinal to shear and then back to longitudinal, occurring at each pair of scattering events. We are interested here in the properties of the "shear" coherent wave that are given by the expansion of its squared wavenumber, around that in the absence of particles, in powers of the concentration. At 1 MHz and a particle radius of 0.05 μm, we show that convergence of the modal series involved in that expansion may be reached after three terms: we use ten terms subsequently. We study the evolution of both the effective shear velocity and attenuation with concentration, as well as that of the effective shear viscosity deduced therefrom.

1. Introduction

We study the propagation of the "shear" coherent wave in a disordered suspension of silica spheres in water. Forrester [1] et al. have shown that for particle radii smaller than a few micrometres the attenuation of the "longitudinal" coherent wave in such a suspension could be well predicted by a multiple scattering model, as long as the shear-mode effects, due to water viscosity, were taken into account, as in Ref. [2]. Including those effects leads to the introduction of another coherent wave, called the "shear" coherent wave (with quotes omitted in the following), because its wavenumber ξ_S reduces to that, k_S, of the shear wave in (viscous) water in the absence of particles. It is in the properties of this shear coherent wave that we are interested here.

The physical properties of both water and silica are the same as in Ref. [1], and the numerical study is conducted at one frequency only, $f = 1$ MHz. All particles have the same radius $a = 0.05$ μm, so that letting k_C denote the compressional wavenumber in water in the absence of particles, while low frequency approximations may be used when dealing with compression waves, one must be careful with shear waves, as $k_C a \simeq 2\times 10^{-4}$, and $k_S a \simeq 0.1\ 1+i$.

The properties of the shear coherent wave are deduced from the calculation of its wavenumber ξ_S, as a function of the concentration (volume fraction) c of solid, using the multiple scattering model of Ref. [2], and section 2 is dedicated to numerical issues that need to be taken care of in doing this. The effective shear velocity and attenuation are studied as a function of the concentration in section 3, and preliminary work on the definition of a prospective effective shear viscosity is conducted in section 4.

2. The effective shear wavenumber

2.1. Effective shear wavenumber expansion

After Eqs. (16,29-32) of Ref. [2], the effective "shear" wavenumber ξ_S may be expanded around k_S in powers of the concentration ($c = 4\pi n_0 a^3/3$),

$$\frac{\xi_{1S}^2}{k_S^2} = 1 + \frac{c}{(k_S a)^3}\delta_1^{SS} \tag{1}$$

$$\frac{\xi_S^2}{k_S^2} \simeq \frac{\xi_{1S}^2}{k_S^2} + \frac{c^2}{(k_S a)^6}\left(\delta_2^{SS} + \delta_2^{SC}\right) + O\left(c^3\right) \tag{2}$$

The first order term (in concentration), δ_1^{SS}, involves a modal series of the form $\sum_{n=0}^{+\infty} f_n T_n^{SS}$, while δ_2^{SS} and δ_2^{SC} involve respectively series of the form $\sum_{n=0}^{+\infty}\sum_{m=0}^{+\infty} g_{nm} T_n^{SS} T_m^{SS}$ and $\sum_{n=0}^{+\infty}\sum_{m=0}^{+\infty} h_{nm} T_n^{SC} T_m^{CS}$, with f_n, g_{nm}, h_{nm} more or less complicated functions of n and m, and the T_n^{pq} (p,q=S,C) the modal scattering coefficients of a single particle for an incident compressional (p=C) or shear (p=S) wave into either a compressional (q=C) or a shear (q=S) one. They are defined in Ref. [2] as associated with the Debye potentials of either compressional or shear waves, with no dependence with the azimuthal angle (see Eqs. 5,6 in Ref. [2]) . Their decay with the increase of mode number n is shown in Figure 1.

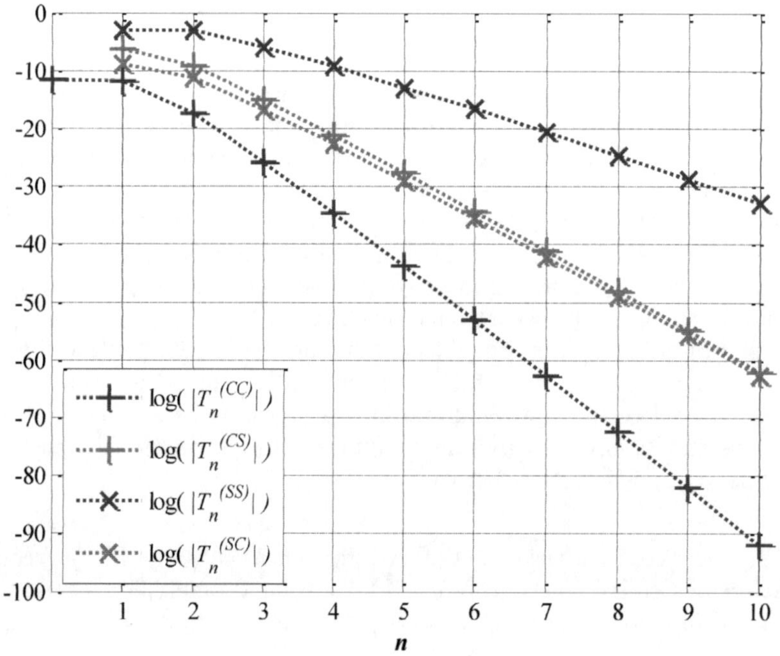

Figure 1. The modal scattering coefficients (amplitudes) as functions of mode number.

The scattering coefficient of a compressional wave into a compressional one is the fastest to decrease, that of a shear wave into a shear one the slowest. The first one is also the lowest in magnitude, while the second one the largest. However, all coefficients are smaller than 10^{-10} as soon as n is larger than 4, and we can suppose that all the series involved in Eqs. (1,2) should have achieved convergence after summing up 5 terms at most.

Convergence of the $\delta_{2N}^{SS} = \sum_{n=0}^{N} \sum_{m=0}^{N} g_{nm} T_n^{SS} T_m^{SS}$ series may be studied through the plot of its real and imaginary parts versus N, as done in Figure 2, where all the δ terms are plotted against N, showing that N=2 should be sufficient for all series to reach convergence. While this information could be useful to carry out analytic approximations of Eqs. (1,2), a value of N=10 has been used nevertheless to obtain all the subsequent results.

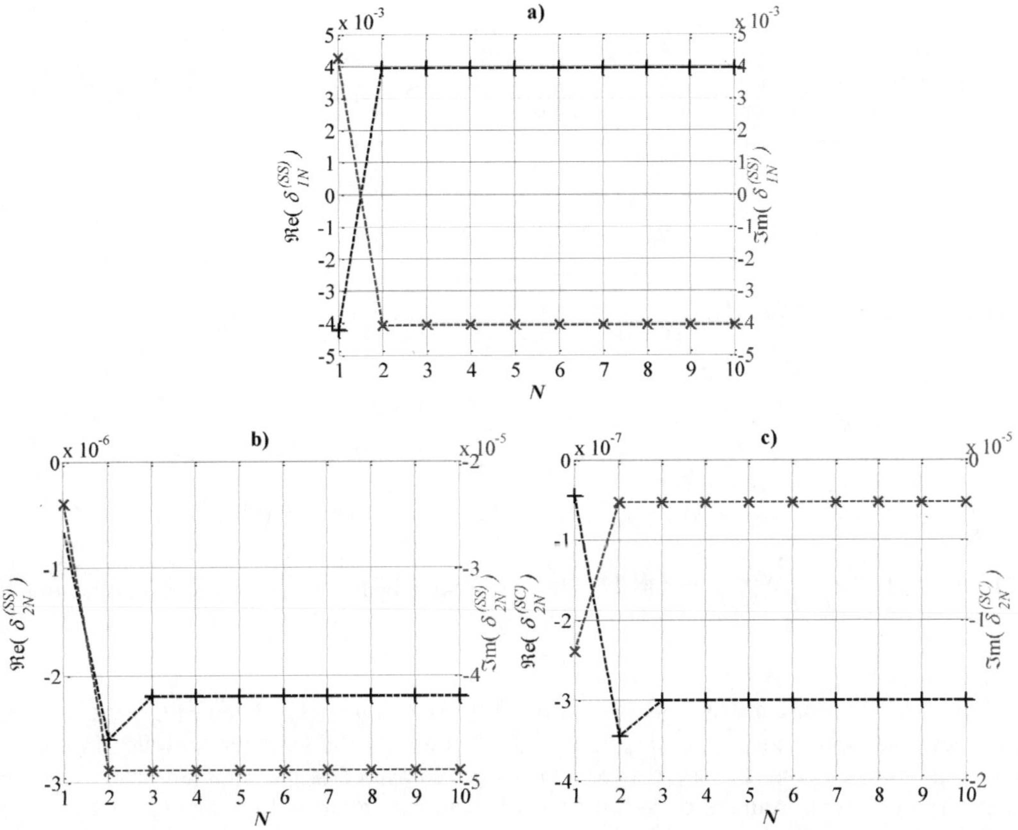

Figure 2. Convergence of the first and second order terms of the effective wavenumber expansion Eqs. (1,2). N is the order at which all modal series are truncated.

Eq. (2) corresponds to the expansion of $\frac{\xi_S^2}{k_S^2}$ in powers of the concentration c, up to order 2. This expansion was carried out [2] under the assumption of a small enough concentration for all successive orders to be decreasing in magnitude. However, each term of a given order r in Eq. (2) is multiplied, not only by the small quantity c^r, as one would have expected, but by a larger one, $\dfrac{c^r}{k_S a^{3r}}$, so that, while the delta terms relating to the terms of order 2 in concentration (Figure 2) are two orders of magnitude smaller than the delta term relating to the order 1 term in concentration, Figure 3 shows that the whole term of order 2 in concentration in Eq. (2) is no longer smaller than the whole term of order 1 in concentration as soon as the concentration gets higher than about 0.25. This shows clearly the limit of the model, and the subsequent analysis considers only concentration up to 0.2.

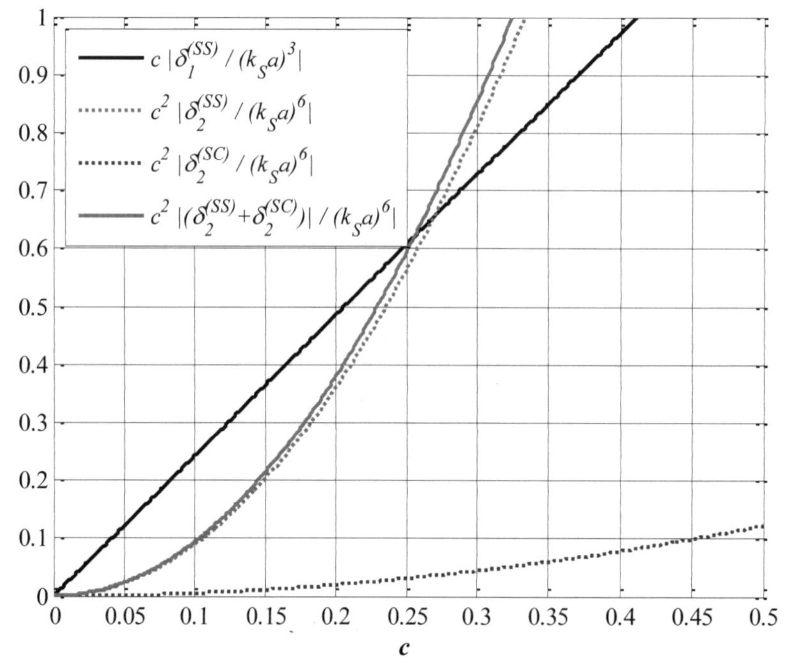

Figure 3. Magnitude of the terms in the effective shear wavenumber expansion in concentration.

2.2. Properties of the effective shear wave

The effect of the concentration of scatterers on the velocity and the attenuation of the effective shear wave is shown in Figure 4. Not surprisingly, the introduction of solid scatterers in water increases the shear velocity and decreases the shear wave attenuation, compared to the situation in the absence of scatterers. The deviation from the corresponding values in the absence of scatterers, however, is less than 10 %. As the effective wavenumber expansion in concentration is limited to order 2 in concentration, both the velocity and attenuation curves exhibit a quadratic behavior, but the concentration c_{crit} at which the velocity reaches a maximum is slightly lower than that at which the attenuation is lowest. As the quadratic behavior of the curves is the most obvious for concentration values around c_{crit}, it might be an indication that, for this particular frequency under study, order 3 in concentration can no longer be neglected for $c \geq c_{crit}$.

Figure 5 shows that the real and imaginary parts of the effective shear wavenumber are practically equal, just like those in the absence of scatterers, and the next section discusses the possibility of defining an effective shear viscosity by analogy with a pure viscous fluid.

IOP Conf. Series: Journal of Physics: Conf. Series **1017** (2018) 012003 doi:10.1088/1742-6596/1017/1/012003

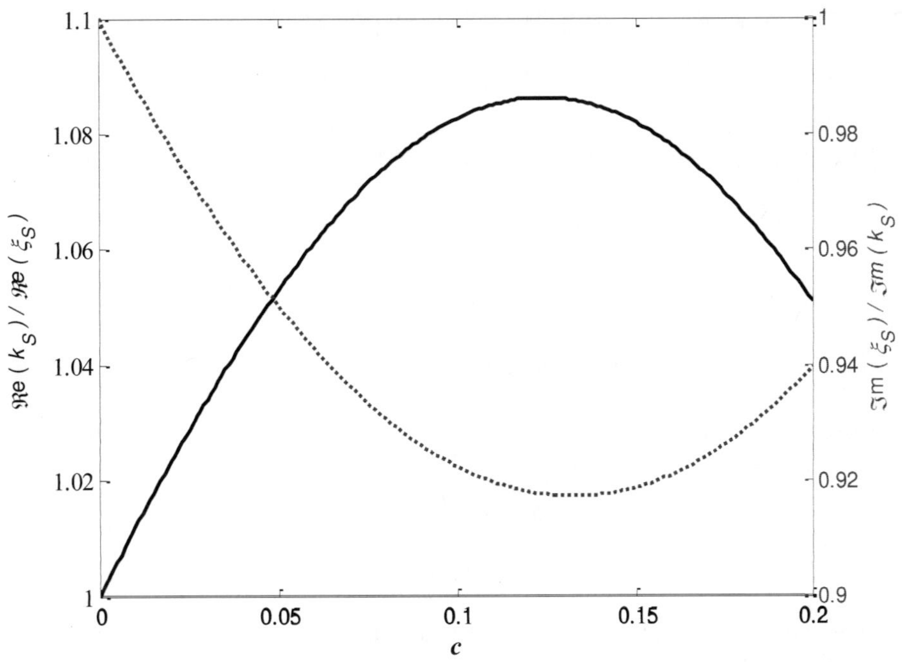

Figure 4. Ratios of the effective shear wave velocity (solid line, left vertical axis) and attenuation (dotted line, right vertical axis) to those of the shear wave in the absence of scatterers, versus concentration.

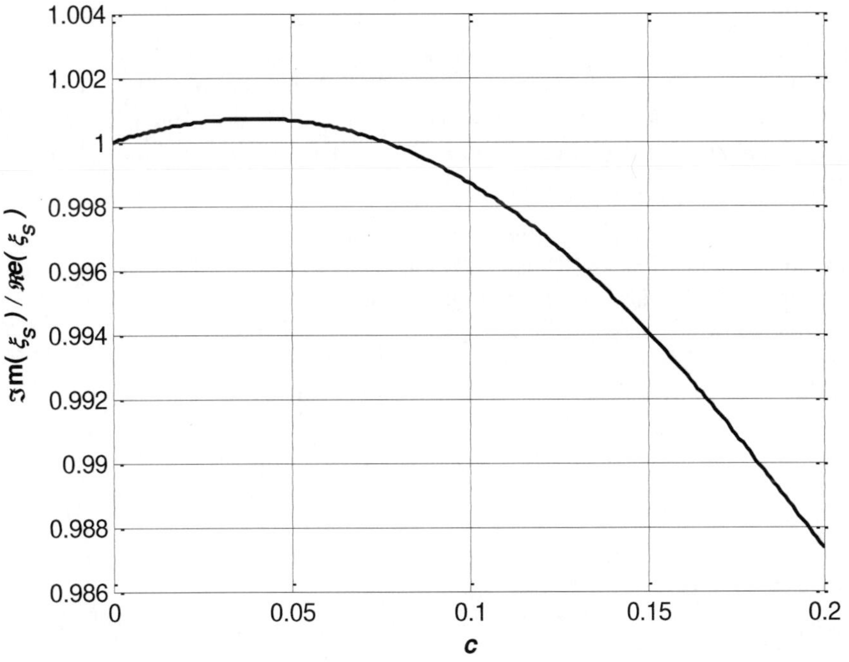

Figure 5. The effective shear wavenumber: ratio of its imaginary part to its real part versus concentration.

IOP Conf. Series: Journal of Physics: Conf. Series **1017** (2018) 012003 doi:10.1088/1742-6596/1017/1/012003

3. Towards the definition of an effective shear viscosity

In a viscous fluid such as water, and supposing the exp($-i\omega t$) time dependence,

$$k_S^2 = \frac{\rho}{\mu}\omega^2, \qquad \mu = -i\omega\eta, \tag{3}$$

so that the shear wavenumber has equal real and imaginary parts. The fact that the effective shear wavenumber calculated in the preceding section almost satisfies this condition suggests that one could define an effective medium, at least from the shear wave point of view, with a real effective mass density ρ_{eff} and a (practically) real effective shear viscosity η_{eff} obeying

$$\frac{\eta}{\eta_{\text{eff}}} = \frac{\xi_S^2}{k_S^2}\frac{\rho}{\rho_{\text{eff}}}. \tag{4}$$

There have been many published works regarding the effective density of a random configuration of scatterers in a host matrix, with nearly as many formulas for that density. All agree on the complex nature of the effective mass density, as well as on its real static limit, which is given [4,5] by a simple volume average in the case of an elastic matrix,

$$\rho_{\text{eff}} = (1-c)\rho_{\text{water}} + c\rho_{\text{Silica}}, \tag{5}$$

and by a different one [4] in the case of an inviscid fluid matrix.

Taking into account the shear viscosity of water through Eq. (3) is equivalent to considering water as a solid with a purely imaginary second (shear) Lamé coefficient μ, and thus adopting Eq. (5) with Eqs. (2, 4) allows an effective viscosity, η_{eff}, to be obtained. This effective viscosity η_{eff}, which is almost real, is plotted in Figure 6 (black curve), and has an imaginary part of less than 2 % of the viscosity of pure water η. The other curves in Figure 6 correspond to a few hydrodynamic models of viscosity for hard sphere suspensions in water that are described in Ref. [5]. While all hydrodynamic models provide quite different results, they all show a monotonic increase of the effective viscosity with concentration, which is not the case in our calculation (black curve). This dramatically different behavior may be explained from the difference in the assumptions of the hydrodynamic models and the one used here. All of them correspond to static expressions of the effective mass density and viscosity, whereas our calculation, while using a static approximation for the effective density (Eq. (5)), has been obtained from a multiple scattering model at intermediate frequency $k_C a \simeq 2\times10^{-4}$, and $k_S a \simeq 0.1\ 1+i$. It is quite clear that one should use here a frequency dependent approximation of the effective mass density, but this is a topic of further investigation. A comparison of our model calculations of the effective compressional and shear wavenumbers with those obtained from various self consistent models such as [6,7] is also an area for future work.

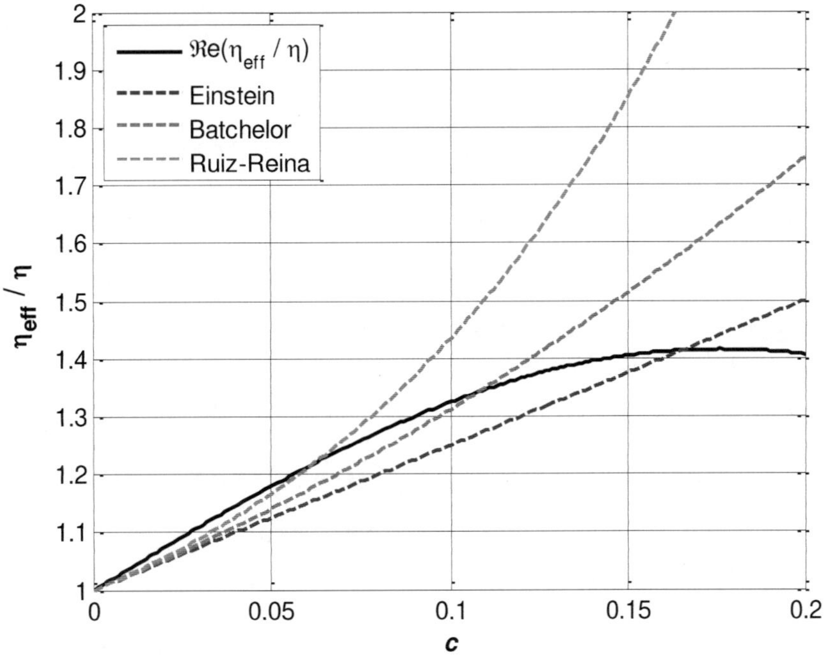

Figure 6. The effective shear viscosity as a function of concentration. Hydrodynamic models follow Ref. [5].

4. Conclusion

This work is a preliminary study of the properties of the shear coherent wave in suspensions of spherical particles. It highlights the truncation of the series in partial wave orders as well as the problem with convergence of the concentration series. Although an effective viscosity has been determined, its validity requires use of a frequency dependent effective mass density. Further work will include the introduction of self consistent models such as in Refs. [6, 7] and their relation with the model used here. The relation between hydrodynamic models and multiple scattering effective theories should be also investigated further, as begun by the authors of Ref. [8].

References

[1] Forrester D M, Huang J, Pinfield V J and Luppé F 2016 *Nanoscale* **8** 5497-5506
[2] Luppé F, Conoir J M and Norris A N 2012 *J. Acoust. Soc. Am.* **131** 1113-20
[3] Pinfield V J 2007 *J. Acoust. Soc. Am.* **122** 205-21
[4] Berryman J G 1980 *J. Acoust. Soc. Am.* **68** 1809-19
[5] Challis R E and Pinfield V J 2014 *Ultrasonics* **54** 1737-44
[6] Mei J, Liu Z, Wen W and Sheng P 2007 Phys. Rev. B 76 134205 1-20
[7] Jin C 2012 *J. Acoust. Soc. Am.* **132** 615-20
[8] Valier-Brasier T, Conoir JM and Coulouvrat F 2015 *J. Acoust. Soc. Am.* **138**, 2598-2612

IOP Conf. Series: Journal of Physics: Conf. Series **1017** (2018) 012004 doi:10.1088/1742-6596/1017/1/012004

Probing flexible thermoplastic thin films on a substrate using ultrasonic waves to retrieve mechanical moduli and density: Inverse problem

H. Lazri[1], E. Ogam[2], B. Amar[1], Z. E. A Fellah[2], N. Sayoud[1] and Y. Boumaiza[1]

[1] Laboratoire d'Elaboration et d'Analyse des Matériaux, Université Badji Mokhtar Annaba, Algeria

[2] Aix Marseille Univ, CNRS, Centrale Marseille, LMA UPR7051, Marseille, France

E-mail: ogam@lma.cnrs-mrs.fr

Abstract. Flexible, supple thermoplastic thin films (PVB and PET) placed on elastic substrates were probed using ultrasonic waves to identify their mechanical moduli and density. The composite medium immersed in a fluid host medium (water) was excited using a 50 Mhz transducer operating at normal incidence in reflection mode. Elastic wave propagation data from the stratified medium was captured in the host medium as scattered field. These data were used along with theoretical fluid-solid interaction forward models for stratified-media developed using elasticity theory, to solve an inverse problem for the recovery of the model parameters of the thin films. Two configurations were modeled, one considering the substrate as a semi-infinite elastic medium and the second the substrate having a finite thickness and flanked by a semi-infinite host medium. Transverse slip for the sliding interface between the films and substrate was chosen. This was found to agree with the experiments whereby the thin films were just placed on the substrate without bonding. The inverse problems for the recovery of the mechanical parameters were successful in retrieving the thin films' parameters under the slip boundary condition. The possible improvements to the new method for the characterization of thin films are discussed.

1. Introduction

This study reports the development of new methods for the characterization of two types of thin flexible films made from thermoplastic polymers by solving an inverse problem using real ultrasound reflection data and theoretical interaction models. The aim of this study is to:

(i) develop new methods/techniques and corresponding geometric configuration for the characterization of flexible, supple thermoplastic thin films (PVB and PET),

(ii) Identify their mechanical moduli (Youngs modulus, Poisson coefficient) and density.

The knowledge of mechanical properties is necessary to understand mechanical behavior and improve the design of thermoplastic thin films.

Thin flexible thermoplastic polymer resin films are employed in a variety of applications

- Thermoplastic thin made from polyEthylene terephthalate (PET) [1] are employed for the reinforcement of glass windscreens.

- Polyvinyl butyral (PVB) is used as protective interlayer films that help hold shattered glass windows together during severe weather conditions, accidental impact or vandalism.

The popular methods for the characterization of thin films are based on scanning acoustic microscopy [2]. These methods have mainly allowed access to the measured velocity of sound waves propagating in the thin films and their V(z) curves [3] at a frequency range from 170-450 MHz. Lavrentyev et al [4] proposed a method for determining the properties of a coating on a thin plate. The method allows simultaneous determination of the coating thickness, density, elastic moduli and attenuation from both the normal and oblique incidences reflection *frequency spectra*. Phase-shift shadow Moiré optical methods have also been employed to recover Young's moduli of *composite* systems composed of various epoxy coated polyethylene terephthalate (PET) micro-cantilevers [5].

2. Materials and methods

2.1. Geometry of the experimental configuration and probing frequency

The films to be characterized were placed on elastic substrates and were then probed using ultrasonic waves to identify their mechanical moduli and density.

The geometry of the acoustic wave reflection configuration by the layered media composed of the thin film (Ω_1) placed on an elastic substrate (Ω_2) and immersed in water (Ω_0) are depicted on Fig. (1). The mode conversions are also represented on the figure. The geometry is represented with an oblique angle of incidence ($\theta \neq 0$) for the acoustic wave impinging on the film. The measurements were done at normal incidence ($\theta = 0$) only. The theoretical models herein were developed using elasticity theory (starting from Hooke's law up to the derivation of the elastic wave propagation equations in the elastic solid as in reference [6]). The general case of an oblique angle of incidence θ was chosen for the model.

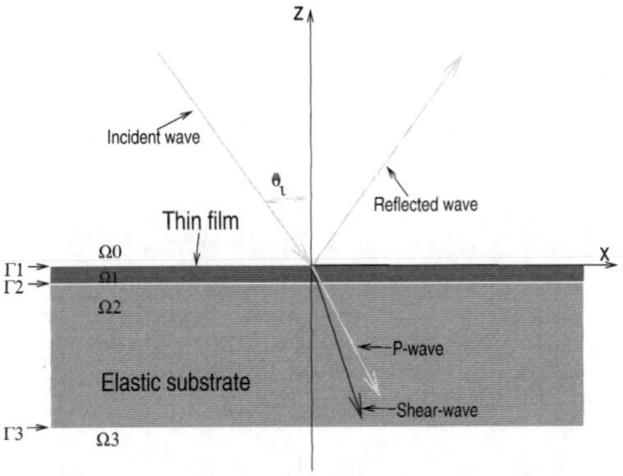

Figure 1. The geometry of the reflection configuration of ultrasonic waves impinging on a thin film (domain Ω_1) placed on an elastic substrate (Ω_2). The semi-infinite fluid media (water) domain is depicted by Ω_0. The boundary between the fluid and the film is Γ_1, that with the substrate Γ_2 and that between the substrate and the fluid Ω_3, is Γ_3. The mode conversions of the incident wave from P-wave into P-wave and shear waves in the thin film (Ω_1) are not shown but are indicated in the substrate.

The probing frequency is imposed by the film thickness and the number of desired modes propagating in the film, $f = nv_L/2L$ (n is the mode number of longitudinal modes, v_L is the wave velocity, L is the thickness of the film).

2.2. The experimental film-substrate configurations studied
Two configurations of the substrate were studied, in which

(i) the film was placed on a substrate with the latter considered as a semi-infinite half space i.e no reflected wave from the second interface Γ_3 of the elastic medium substrate,
(ii) the substrate is of finite thickness (infinite elastic plate) with the composite structure (film on the substrate) bounded on both sides by an infinite fluid medium. A pressure wave is transmitted at the second interface of the substrate Γ_3

The film and the substrate were not glued together. The film was allowed to slide on the substrate.

3. Modeling the configurations
3.1. The domain equations - frequency domain
The plane wave harmonic solutions of the wave equations for each elastic domains are written using potentials. The acoustic pressure wave in the fluid domain is,

$$p_f(x,z) = \left(\mathrm{e}^{\kappa z_f z} + R\mathrm{e}^{-\kappa z_f z}\right)\mathrm{e}^{\kappa_f x} p_i \quad z \in \Omega_f, \tag{1}$$

where R is the reflection coefficient, p_i is the amplitude of the incident wave in the fluid, the complex wavenumber in the fluid $\kappa z_f = \mathrm{i}k_f \cos(\theta_i)$ ($k_f = \omega/c_f$ and c_f is the wave velocity in the fluid) and the subscript f stand for fluid.

In the film layer, the potentials of elastic waves have the following form (for the incident and reflected waves),

$$\begin{aligned}\Psi^{s1}(x,z) &= \left(\Phi_{11}\,\mathrm{e}^{\kappa z_{11}\,z} + \Phi_{12}\,\mathrm{e}^{-\kappa z_{11}\,z}\right)\mathrm{e}^{\kappa_f x} \quad z \in \Omega_1,\\ &= \left(A_{11}\cosh(\kappa z_{11}\,z) + A_{12}\sinh(\kappa z_{11}\,z)\right)\mathrm{e}^{\kappa_f x} \quad z \in \Omega_1,\end{aligned} \tag{2}$$

where the compressional wave potential amplitudes $A_{11} = \Phi_{11} + \Phi_{12}$, $A_{12} = \Phi_{11} - \Phi_{12}$ and A_{11} and A_{12}) (after conversion of the exponential expressions into trigonometrical functions). The superscript s stands for solid.

$$\begin{aligned}\Phi^{s1}(x,z) &= \left(B_{11}\cosh(\kappa z_{31}\,z) + B_{12}\sinh(\kappa z_{31}\,z)\right)e^{\kappa_f x} \quad z \in \Omega_1,\\ \Psi^{s2}(x,z) &= \Phi_{21}\,\mathrm{e}^{\kappa z_{12}\,(z+L)}\mathrm{e}^{\kappa_f x} \quad z \in \Omega_2,\\ \Phi^{s2}(x,z) &= \Psi_{21}\,\mathrm{e}^{\kappa z_{32}\,(z+L)}\mathrm{e}^{\kappa_f x} \quad z \in \Omega_2,\end{aligned} \tag{3}$$

where the amplitudes of the shear waves are $B_{11} = \Psi_{11} + \Psi_{12}$ and $B_{12} = \Psi_{11} - \Psi_{12}$.

The amplitudes Φ_{11} and Ψ_{11} are for the incident P-wave and shear-wave in the thin film respectively, while Φ_{12} and Ψ_{12} are amplitudes of the reflected P-wave and shear-wave respectively and L is the thickness of the film, Φ_{21} and Ψ_{21} are the amplitudes of the incident P-wave and shear-wave in the substrate respectively

3.2. The complex wavenumbers
The complex wave-numbers $\kappa_f = \mathrm{i}\gamma_f$ ($\mathrm{i} = \sqrt{-1}$), $\gamma_f = k_f sin(\theta)$ and the compressional complex wave numbers κz_{1n} and the shear ones κz_{3n} ($n = 1, 2$ represents the medium Ω_n) are given by

$$\begin{aligned}\kappa z_{11} &= \mathrm{i}\sqrt{k_{11}^2 - \gamma_f^2}, \quad \kappa z_{12} = \mathrm{i}\sqrt{k_{12}^2 - \gamma_f^2},\\ \kappa z_{31} &= \mathrm{i}\sqrt{k_{31}^2 - \gamma_f^2}, \quad \kappa z_{32} = \mathrm{i}\sqrt{k_{32}^2 - \gamma_f^2}.\end{aligned} \tag{4}$$

3.2.1. The relationship between wave numbers with the elastic constants The wave numbers are given by

$$k_{11} = \omega\sqrt{\frac{\rho_1}{M_1}}, \quad k_{31} = \omega\sqrt{\frac{\rho_1}{\mu_1}}$$
$$k_{12} = \omega\sqrt{\frac{\rho_2}{M_2}}, \quad k_{32} = \omega\sqrt{\frac{\rho_2}{\mu_2}} \tag{5}$$

where ω, angular frequency of the wave, ρ_n , μ_n and M_n are the density, shear modulus and the P-wave modulus of the layer respectively. The inherent damping of the films were taken into account by introducing a scalar-valued loss factor η_n^s into Young's modulus, such that $E_n^s = E_n^s(1 + i\eta_n^s)$ (E_n^s is the dynamic Young's modulus of the layer). The relationships between M_n and E_n^s are given in [6].

3.3. Thin flexible film resting on a substrate and both immersed in a fluid - the boundary conditions (BC)

3.3.1. BC between the thin film and the fluid The thin film is bounded on one side by semi-infinite fluid half space (Ω_0) while the other side is bounded by a semi-infinite elastic half space.

The boundary conditions at the interface Γ_1 between the thin film and the fluid are ; continuity of the velocity and the normal stress while the tangential stress (σ_{xz}) is zero across the interface at $z \in \Gamma_1$

$$v^f(x, 0, s) = v_z^{s1}(x, 0, s), \quad p_f(x, 0, \omega) = \sigma_{zz}^{s1}(x, 0, s), \quad \sigma_{xz}^{s1}(x, 0, s) = 0, \tag{6}$$

where the superscript and subscript s and f stand for solid and fluid respectively.

3.3.2. BC between the thin film and the substrate - Sliding boundary conditions The thin film was placed on the substrate without gluing. The water between them was squeezed out by sliding the edge of a flat ruler over the film. The measurement was taken immediately before water infiltrated between the film and substrate interface (a slow process). The boundary conditions between the thin film and the elastic substrate at the interface Γ_2 was chosen to be that of transverse slip (sliding can occur). Overall, there is; continuity of the normal velocity, the normal stress, while the tangential stress (σ_{xz}) is zero across the interface at $z \in \Gamma_2$

$$v_z^{s1}(x, -L, s) = v_z^{s2}(x, -L, s), \quad \sigma_{zz}^{s1}(x, z, s) = \sigma_{zz}^{s2}(x, -L, s),$$
$$\sigma_{xz}^{s1}(x, -L, s) = \sigma_{xz}^{s2}(x, -L, s) = 0.$$

4. The direct and inverse problem for the retrieval of mechanical parameters of thin films

4.1. The direct problem

The relationships employed for the displacement (u), velocity (v) with the wave potentials in the Laplace domain are

$$\begin{aligned}
u_x^{sn} &= \frac{\partial}{\partial x}\Psi^{sn}(x,z) - \frac{\partial}{\partial z}\Phi^{sn}(x,z), \quad n = 1,2, \\
v_x^{sn} &= s\frac{\partial}{\partial x}\Psi^{sn}(x,z) - s\frac{\partial}{\partial z}\Phi^{sn}(x,z), \\
u_z^{sn} &= \frac{\partial}{\partial z}\Psi^{sn}(x,z) + \frac{\partial}{\partial x}\Phi^{sn}(x,z), \\
v_z^{sn} &= s\frac{\partial}{\partial z}\Psi^{sn}(x,z) + s\frac{\partial}{\partial x}\Phi^{sn}(x,z)
\end{aligned} \tag{7}$$

4.1.1. The mechanical stresses The stresses in the elastic solid medium are given by

$$\sigma_{zz}^{sn} = \lambda_n \left(\frac{\partial}{\partial x} u_x^{sn}(x,z) + \frac{\partial}{\partial z} u_z^{sn}(x,z) \right) + 2\,\mu_n \frac{\partial}{\partial z} u_z^{sn}(x,z), \quad n = 1,2,$$
$$\sigma_{xz}^{sn} = \mu_n \left(\frac{\partial}{\partial z} u_x^{sn}(x,z) + \frac{\partial}{\partial x} u_z^{sn}(x,z) \right), \tag{8}$$

4.2. The matrix equation of the stratified media - film-substrate interface are sliding
The matrix equation was obtained by combining the domain potential equations, the mechanical stress relationships and the boundary equations when the substrate is a semi-infinite half-space and is given by:

$$\begin{bmatrix} 0 & s\kappa z_{11} & s\kappa_f & 0 & 0 & 0 & -\frac{\cos(\theta) p_i}{c_f \rho_f} \\ -\sinh(\kappa z_{11} L)\kappa z_{11}\, s & \cosh(\kappa z_{11} L)\kappa z_{11}\, s & \cosh(\kappa z_{31} L)\kappa_f\, s & -\sinh(\kappa z_{31} L)\kappa_f\, s & -s\kappa z_{12} & -s\kappa_f & 0 \\ \xi_1 & 0 & 0 & 2\,\kappa z_{31}\,\mu_1\,\kappa_f & 0 & 0 & p_i \\ \xi_1 \cosh(\kappa z_{11} L) & -\xi_1 \sinh(\kappa z_{11} L) & -2\,\kappa z_{31}\,\mu_1\,\kappa_f \sinh(\kappa z_{31} L) & 2\cosh(\kappa z_{31} L)\,\mu_1\,\kappa_f\,\kappa z_{31} & -\xi_2 & -2\,\mu_2\,\kappa_f\,\kappa z_{32} & 0 \\ 0 & 2\,\mu_1\,\kappa_f\,\kappa z_{11} & \mu_1\left(\kappa_f{}^2 - \kappa z_{31}{}^2\right) & 0 & 0 & 0 & 0 \\ -2\,\mu_1 \sinh(\kappa z_{11} L)\kappa_f\,\kappa z_{11} & 2\,\mu_1 \cosh(\kappa z_{11} L)\kappa_f\,\kappa z_{11} & \mu_1 \cosh(\kappa z_{31} L)\left(\kappa_f{}^2 - \kappa z_{31}{}^2\right) & -\mu_1\left(\kappa_f{}^2 - \kappa z_{31}{}^2\right)\sinh(\kappa z_{31} L) & 0 & 0 & 0 \\ 0 & 0 & 0 & 0 & 2\,\mu_2\,\kappa_f\,\kappa z_{12} & \mu_2\left(\kappa_f{}^2 - \kappa z_{32}{}^2\right) & 0 \end{bmatrix} \times$$
$$\begin{bmatrix} A_{11} \\ A_{12} \\ A_{21} \\ A_{22} \\ \Phi_{12} \\ \Psi_{12} \\ R^{IM} \end{bmatrix} = \begin{bmatrix} -\frac{\cos(\theta) p_i}{c_f \rho_f} \\ 0 \\ -p_i \\ 0 \\ 0 \\ 0 \\ 0 \end{bmatrix}, \tag{9}$$

where $\xi_1 = \left((2\,\mu_1 + \lambda_1)\,\kappa z_{11}{}^2 + \lambda_1\,\kappa_f{}^2\right)$ and $\xi_2 = (2\,\mu_2 + \lambda_2)\,\kappa z_{12}{}^2 + \lambda_2\,\kappa_f{}^2$.
The matrix equation is a linear system of the form $\mathbf{A}.\mathbf{x} = \mathbf{b}$ and was solved using a linear solver that returns the vector $\mathbf{x}$ that satisfies the equation. The reflection coefficient R^{IM} was then extracted from the solution of this equation. This constitutes the interaction model (IM).

4.3. The inverse problem - construction of the cost functions
The viscoelastic interaction model of the acoustic wave reflection R^{IM} of the setup geometry is obtained from the matrix equation developed in the previous subsection (viscoelastic by the introduction of a complex Young's modulus [6]). An error functional expressing the distance between the experimental data reflection coefficient from that of the interaction model was computed for each set of trial values of the parameters.

$$\mathfrak{J}(E_1, \nu_1, \rho_1, E_2, \nu_2, \rho_2) = \sum_{n=1}^{N_s} \left(|R^{IM}(\omega_n, E_1, \nu_1, \rho_1, E_2, \nu_2, \rho_2)| - |R_{IR}^{\exp}(\omega_n)| \right)^2, \tag{10}$$

where $|R_{IR}^{\exp}|$ is the amplitude of the reflection coefficient of the thin film on its substrate obtained from the tank experimental data and $|R^{IM}|$ is the amplitude of the interaction model reflection coefficient. The method of signal processing for obtaining the reflection coefficient from the measured incident and reflected signals from the thin film by computing the transfer function, are explained in detail in reference [7]. The parameters were retrieved by minimizing the objective functional $\mathfrak{J}$. The substrate parameters E_2, ν_2, ρ_2 were supposed known.
The problem was split into the computation of three cost functions ($\mathfrak{J}_E$, $\mathfrak{J}_\nu$, $\mathfrak{J}_\rho$) where each of the model parameters (E_{1s_m}, ν_{1s_m}, ρ_{1s_m}, where s is the step, $m = 1 \ldots p$ and p is the number of intervals) was varied while the other two were fixed. This was done in various intervals not too wide , e.g. Poisson ratio interval was chosen from 0.1 to 0.5, density from 800 to 2000, etc.
Three initial values of the sought-for film parameters (E_{1si}, ν_{1si}, ρ_{1si}) were chosen at the initial step. Some of these values, found in the literature (e.g for PET [8, 9], for both PET and PVB [10])

were obtained using different methods. The solution for each of the parameters was found from the position of the global minimum of the cost function (CF) exhibiting a parabolic shape. The values corresponding to the three CF minima constituted the solution to the inverse problem for the step. The three new minima values were then used to update the initial values for the next step. The search for the optimized values was a manual iterative process repeated for each parameter at a time, at each iteration/step, until convergence was achieved for all the three sought for parameters.
The objective functional curves for each parameter were then reported after assuring that all the three parameters had converged to their optimal values. The curves were plotted with respect to one of the three problem variables, the others being fixed at their computationally-optimal values, i.e, as 2-D plots [11].

5. Results

The captured temporal incident and reflected signals and their calculated spectra are presented in Fig. (2). A correction was undertaken by shifting the signals so as to assure artificially that the transducer-film and transducer-reflector distances were the same. The opposite face of the substrate was used as mirror to obtain the incident signal. The spectrum of the reflected (incident) signal shows the useful bandwidth.

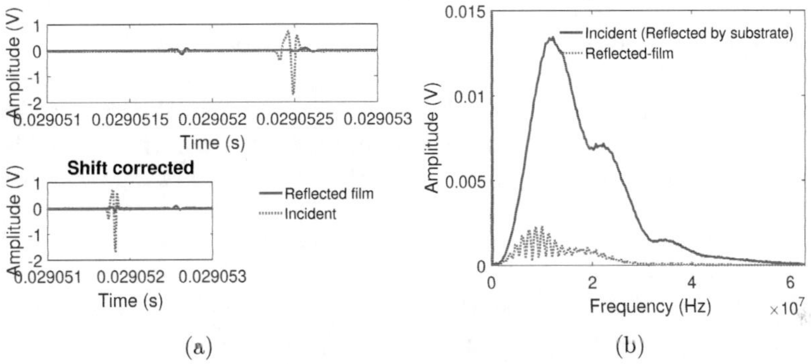

Figure 2. (a) The reflected temporal signals, lower panel, the shifted signals (b) Their frequency spectra for the 750 μm thick PVB film.

5.1. The substrate is a semi-infinite half-space elastic medium

5.1.1. Retrieval of mechanical parameters - cost functions The cost functions are computed using relation in Eqn. (10). They are depicted in Fig. 3.

Measurements of the reflections coefficients were done for a 100μm PET film placed on substrates made of different elastic materials and geometries (glass plate, aluminium slab, aluminium cylinder). This was to analyze the influence of different elastic materials on the reflection coefficient.

5.2. Summary of the results

The results are summarized in Table (1). The densities were compared with those obtained from small specimens of the film whose weights were measured on an analytic electronic balance (Fisher scientific model PAS214C, Pittsburgh, PA USA). The velocities were measured using the time of flight and then compared with those computed using the recovered parameters. The

IOP Conf. Series: Journal of Physics: Conf. Series **1017** (2018) 012004 doi:10.1088/1742-6596/1017/1/012004

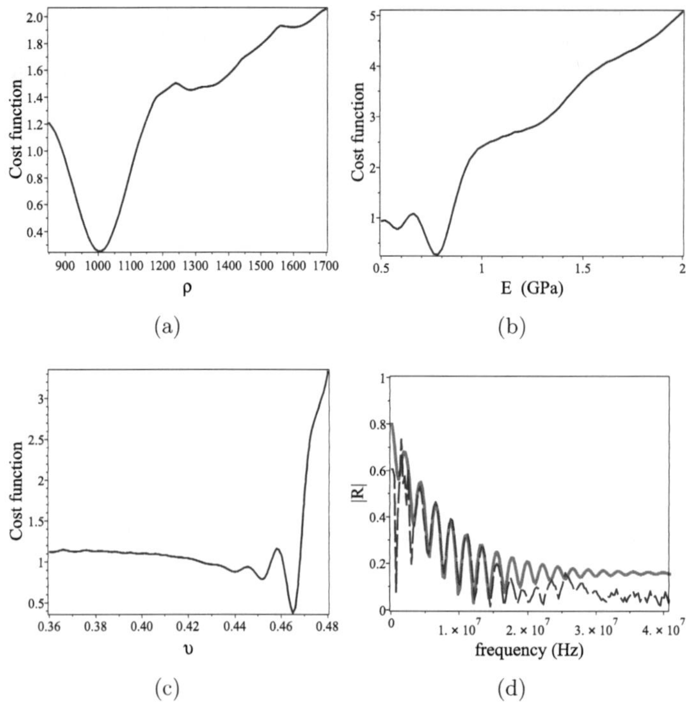

Figure 3. The cost functions for the recovery of the three mechanical parameters for a 450 μm PVB film (a) the density ρ (kg/m^3), (b) Young's modulus, (c) Poisson ratio. (d) The comparison theoretical/experimental reflection coefficients using the recovered parameters.

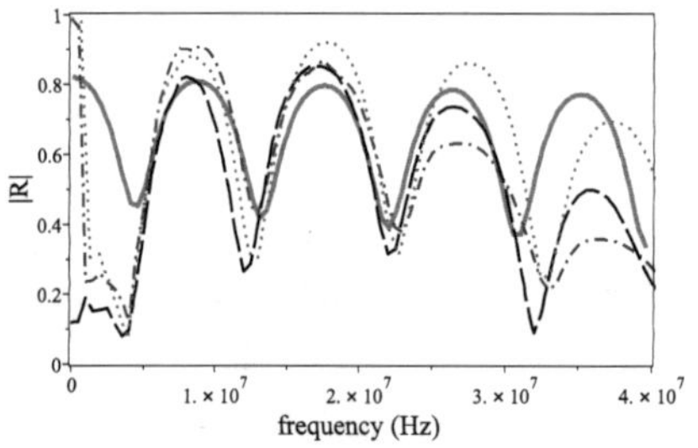

Figure 4. The comparison between the model (solid line) and the measured reflection coefficients for the 100μm PET film resting on the glass plate substrate (dotted line), aluminium slab substrate (long-dash) and aluminium cylinder (dash-dot), with the interface is considered as sliding.

densities were recovered from the electronic balance measurements results were within 3 percent difference with the ones recovered by solving the inverse problems.

The apparent dispersion of properties for different thicknesses of the PVB in Table 1 stems from the structural differences with the thickness of the film as shown in Fig. (5) of the Scanning electron microscopy (SEM) images of the PVB films of different thicknesses.
X-ray diffraction (XRD) and reflectivity methods were also employed to reveal the density as an absolute value and also the surface and interface roughness of the PVBs but whereas these results are yet to be exploited, they showed the same dispersion in the density with the thickness

Table 1. Summary of the identified mechanical parameters with the measured velocity recovered from the time of flight of the longitudinal wave in the film layer.

	Thickness (μm)	Velocity m/s measured	C_{L1}	Young's modulus GPa	Poisson ratio	Loss η_1^s	Density (kg/m^3)
	750	1904	1850	0.8	0.455	0.085	980
PVB	450	1991	1979	0.77	0.465	0.095	1015
	380	1603	1581	0.8	0.45	0.08	1217
PET	100	1727	1788	1.175	0.456	0.025	1560

variation as those shown in the table.
In order to validate the values using the same specimens, the mechanical behaviour of the PVB films will be investigated in future using the small-strain behaviour through dynamic mechanical analysis (DMA) at frequencies from 1 to 100 Hz.

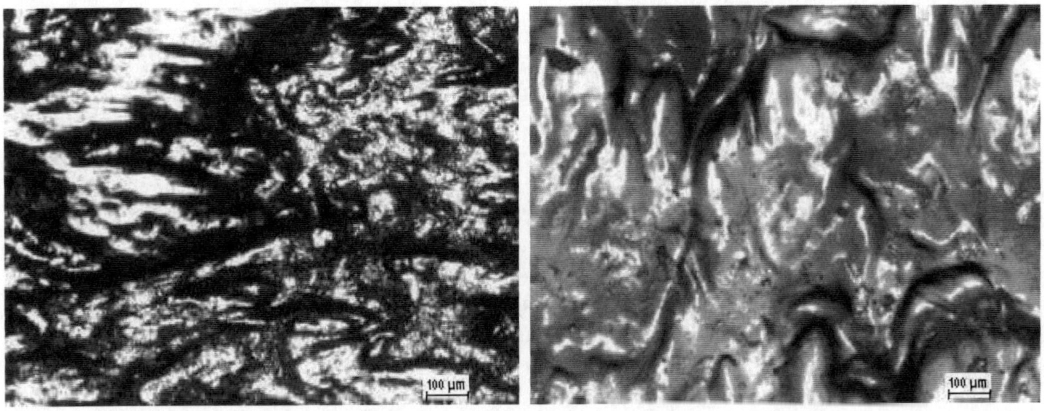

(a) (b)

Figure 5. Scanning electron microscopy images of the PVB films of thickness (a) 450 μm (b) 750 μm

6. Improved modeling - substrate of finite thickness

In this case the substrate is of finite thickness (L_2)and the stratified media is considered as bounded on both sides by the fluid. The procedure is the same as that of the substrate considered as a semi-infinite half-space. The changes in the domain potentials are only for the substrate and the presence of a transmitted pressure wave in the fluid ($z \leq -(L + L_2)$):

$$\begin{aligned} \Psi^{s2}(x,z) &= \left(A_{21}\cosh\left(\kappa z_{12}\ (z+L)\right) + A_{22}\cosh\left(\kappa z_{12}\ (z+L)\right)\right) e^{\kappa_f x} \quad z \in \Omega_2, \\ \Phi^{s2}(x,z) &= \left(B_{21}\cosh\left(\kappa z_{32}\ (z+L)\right) + B_{21}\cosh\left(\kappa z_{32}\ (z+L)\right)\right) e^{\kappa_f x} \quad z \in \Omega_2, \\ p_f^T(x,z) &= e^{(\kappa z_{11} L + \kappa z_{12} L_2) + \kappa z_f (L+L_2+z)} T e^{\kappa_f x}\, p_i \quad z \leq -(L+L_2), \end{aligned} \tag{11}$$

where L_2 is the thickness of the substrate, T is the transmission coefficient of the stratified medium. A matrix equation is obtained as previously and the reflection coefficient computed.

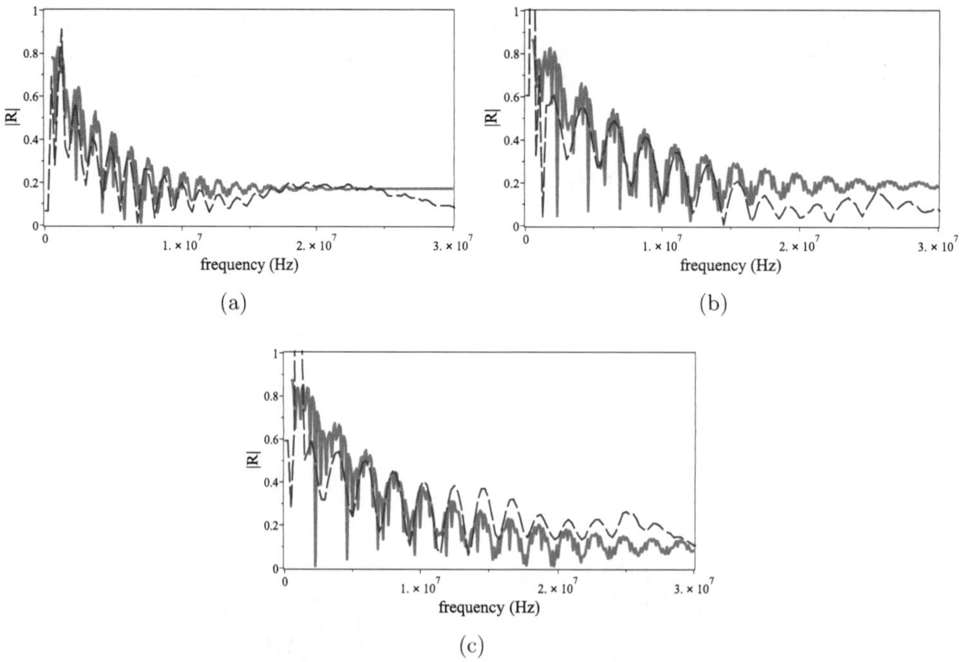

Figure 6. The comparison between the theoretical model (continous line) and the measured (dash) reflection coefficients of thin PVB film placed on the glass substrate with the finite thickness and bordered by a semi-infinite fluid half space (a) 750μm (b) 450μm (c) 380μm.

When the measurement window captures the temporal reflected waves from the face at $z = -(L + L_2)$ the reflection coefficient is perturbed as the computed ones in Fig. (6) but the modes do not change.

7. Conclusion

A method for the recovery of the mechanical parameters of flexible thermoplastic thin films (PETs and PVBs) on elastic substrates using reflected ultrasonic acoustic waves and interaction models has been developed. Two models were developed in this study, one in which the substrate was considered as a semi-infinite half-space and in the second one it had a finite thickness. In both configurations the sliding boundary contact condition between the thin film and substrate were considered as sliding. The stratified medium was immersed in a water tank where it was bounded on both sides by semi-infinite half spaces. The results of the recovered parameters agree well with those in the literature.

References

[1] Toufaili F A E 2006 *Catalytic and Mechanistic Studies of Polyethylene Terephthalate Synthesis* Ph.D. thesis Technische Universität Berlin URL `https://doi.org/10.14279/depositonce-1333`

[2] Briggs A and Kolosov O 2009 *Acoustic Microscopy* (Oxford University Press) ISBN ISBN-13:9780199232734 URL `http://dx.doi.org/10.1093/acprof:oso/9780199232734.001.0001`

[3] Maebayashi M, Matsuoka T, Koda S, Hashitani R, Nishio T and ichi Kimura S 2004 *Polymer* **45** 7563 – 7569 ISSN 0032-3861 URL `http://www.sciencedirect.com/science/article/pii/S0032386104008705`

[4] Lavrentyev A I and Rokhlin S I 2001 *Ultrasonics* **39** 211 – 221 ISSN 0041-624X URL `http://www.sciencedirect.com/science/article/pii/S0041624X00000664`

[5] Lim J, Ratnam M, Azid I and Mutharasu D 2011 *Optics and Lasers in Engineering* **49** 1301 – 1308 ISSN 0143-8166 URL `http://www.sciencedirect.com/science/article/pii/S0143816611001710`

IOP Conf. Series: Journal of Physics: Conf. Series **1017** (2018) 012004 doi:10.1088/1742-6596/1017/1/012004

[6] Ogam E, Fellah Z and Ogam G 2016 *Composite Structures* **135** 205 – 216 ISSN 0263-8223 URL http://www.sciencedirect.com/science/article/pii/S026382231500865X

[7] Ben Mansour M, Ogam E, Fellah Z E A, Soukaina Cherif A, Jelidi A and Ben Jabrallah S 2016 *J. Acoust. Soc. Am.* **139** 2551–2560 URL http://scitation.aip.org/content/asa/journal/jasa/139/5/10.1121/1.4948573

[8] Li T C, Wu B H and Lin J F 2011 *Thin Solid Films* **519** 7875 – 7882 ISSN 0040-6090 URL http://www.sciencedirect.com/science/article/pii/S0040609011005682

[9] Bistričić L, Borjanović V, Leskovac M, Mikac L, McGuire G E, Shenderova O and Nunn N 2015 *Journal of Polymer Research* **22** 39 ISSN 1572-8935 URL http://dx.doi.org/10.1007/s10965-015-0680-z

[10] Shim G I, Kim S H, Ahn D L, Park J K, Jin D H, Chung D T and Choi S Y 2016 *Composites Part B: Engineering* **97** 150 – 161 ISSN 1359-8368 URL http://www.sciencedirect.com/science/article/pii/S1359836816305571

[11] Ogam E, Fellah Z, Sebaa N and Groby J P 2011 *J. Sound Vibr.* **330** 1074 – 1090

IOP Conf. Series: Journal of Physics: Conf. Series **1017** (2018) 012005 doi:10.1088/1742-6596/1017/1/012005

Development of methods for the analysis of multi-mode TFM images

Sy K[1,2,3], Bredif P[2], Iakovleva E[2], Roy O[1] and Lesselier D[3]

[1] M2M, 1, rue de Terre Neuve, 91940 Les Ulis, France
[2] CEA/LIST/DISC Centre de Saclay 91191 Gif-sur-Yvette cedex, France
[3] Laboratoire des Signaux et Systèmes, CNRS-CentraleSupélec-Univ. Paris-Sud, Université Paris Saclay, 3, rue Joliot-Curie, 91192 Gif-sur-Yvette; France

E-mail: kombosse.sy@cea.fr,k.sy@m2m-ndt.com, philippe.bredif@cea.fr, ekaterina.iakovelva@cea.fr, dominique.lesselier@l2s.centralesupelec.fr

Abstract. TFM (Total Focusing Method) is an advanced post-processing imaging algorithm of ultrasonic array data that shows good potential in defect detection and characterization. It can be employed using an infinite number of paths between transducer and focusing point. Depending upon the geometry and the characteristics of the defect in a given part, there are not the same modes that are appropriate for the defect reconstruction. Furthermore, non-physical indications can be observed, prone to misinterpretation. These imaging artifacts are due to the coexistence of several contributions involving several modes of propagation and interactions with possible defects and/or the geometry of the part. Two methods for filtering artifacts and reducing the number of TFM images are developed and illustrated.

1. Introduction

TFM imaging (Total Focusing Method, [1]-[4]) shows a real interest in non-destructive testing since it enables to get an image of the defect facilitating the interpretation of the results. Based on ray tracing, it allows to account for a large number of paths between the transducer and the focusing point: this is the multi-mode TFM imaging ([4][5]).

Depending upon the considered sound path, two kinds of echoes may be observed in the case of a crack: tips diffraction, corresponding to the lower and the upper limits of the crack; and specular echo over the entire length of the defect, which provides an image of the defect. This signature helps to know about the nature of the defect, and thereby eases the inspection. However, the coexistence of several propagation modes yields a large number of images, rendering their analysis difficult, and in addition some indications can be mispositioned within the images, which may lead to further misinterpretation.

This contribution shows how to possibly improve multi-mode TFM images and facilitate their analysis. This paper is organized as follows. In the first section, we recall the principle of multimode TFM imaging. In the second section we illustrate its sensitivity to parameters of interest. Then, in the two last sections, we propose two methods, geometry artifacts filtering and fusion of images, for improving multi-mode TFM imaging.

2. FMC and TFM imaging

Full Matrix Capture (FMC) [6] is a method of data acquisition. It enables to retrieve full transient ultrasonic signals for each transmit and receive element of a linear array probe. For a transducer made of N elements, it consists in N successive shots wherein each element transmits a signal and all receive. So a matrix $S(t)$, the multi-static matrix, of N by N elementary signals is obtained, and the TFM post-processing can be applied to it.

The imaging method itself is based on the principle of delay and sum. Focusing is achieved at each point of a pre-defined region of interest, and the amplitude $I^m(P)$ in each pixel P for the mode m is as

$$I^m(P) = \sum_{i=1}^{N}\sum_{j=1}^{N} S_{ij}(t_{ij}^P) \tag{1}$$

wherein t_{ij}^P is the time of flight between transmitter i, receiver j, and focusing point P.

There exists a large number of paths or modes between the transducer and any given point of focusing increased by the coexistence of longitudinal (L) and transverse (T) waves and the possible conversion of waves at the interfaces and on the defect. Three paths or modes are commonly used [7]: direct modes (LL, LT, TL, TT), where the path is without any rebound on the interfaces; corner echo modes (half skip Total Focusing Method) (LLL, LLT, LTL, ...), where the transmitted wave bounces on the bottom before interacting with the defect; and the indirect modes (LLLL, LLLT, LLTL, ...), where the wave bounces on the bottom before and after interacting with the defect.

However, not all such modes are useful. Indeed some provide pertinent information on the defect, yet some just bring artifacts, thereupon rendering difficult the analysis of TFM images. In the next section, the sensitivity of the TFM imaging to some parameters is considered in more detail.

3. Sensitivity of TFM imaging to defect characteristics

TFM images obtained for a given configuration vary depending upon the modes of reconstruction used. Indeed various observations have shown that the TFM imaging is very sensitive to the defect characteristics as well as to the geometry of the part to be inspected.

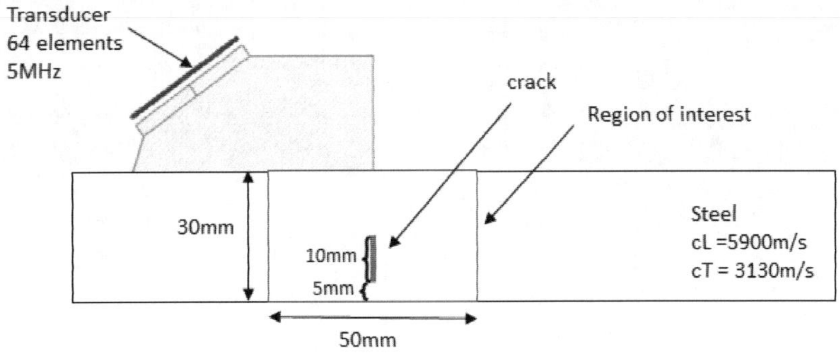

Figure 1: Configuration of inspection.

To investigate the influence of parameters, a configuration of inspection has been defined, sketched in Figure 1. The structure to be inspected is a part with plane and parallel sides which is made of homogeneous steel with a 30mm depth. The inspection is performed with a transducer of 64 elements, the central frequency of which is 5MHz, and the pitch is 0.6mm. We

IOP Conf. Series: Journal of Physics: Conf. Series **1017** (2018) 012005 doi:10.1088/1742-6596/1017/1/012005

intend to observe the influence of defect characteristics onto the TFM imaging as a function of the mode used.

In this case, only the corner echo modes, which are allowing specular reflection for the considered defects, are used.

3.1. One defect, several images

A crack of 10 mm height whose end is 5mm from the bottom is retrieved with using 8 corner echo modes, refer to Figure 4.

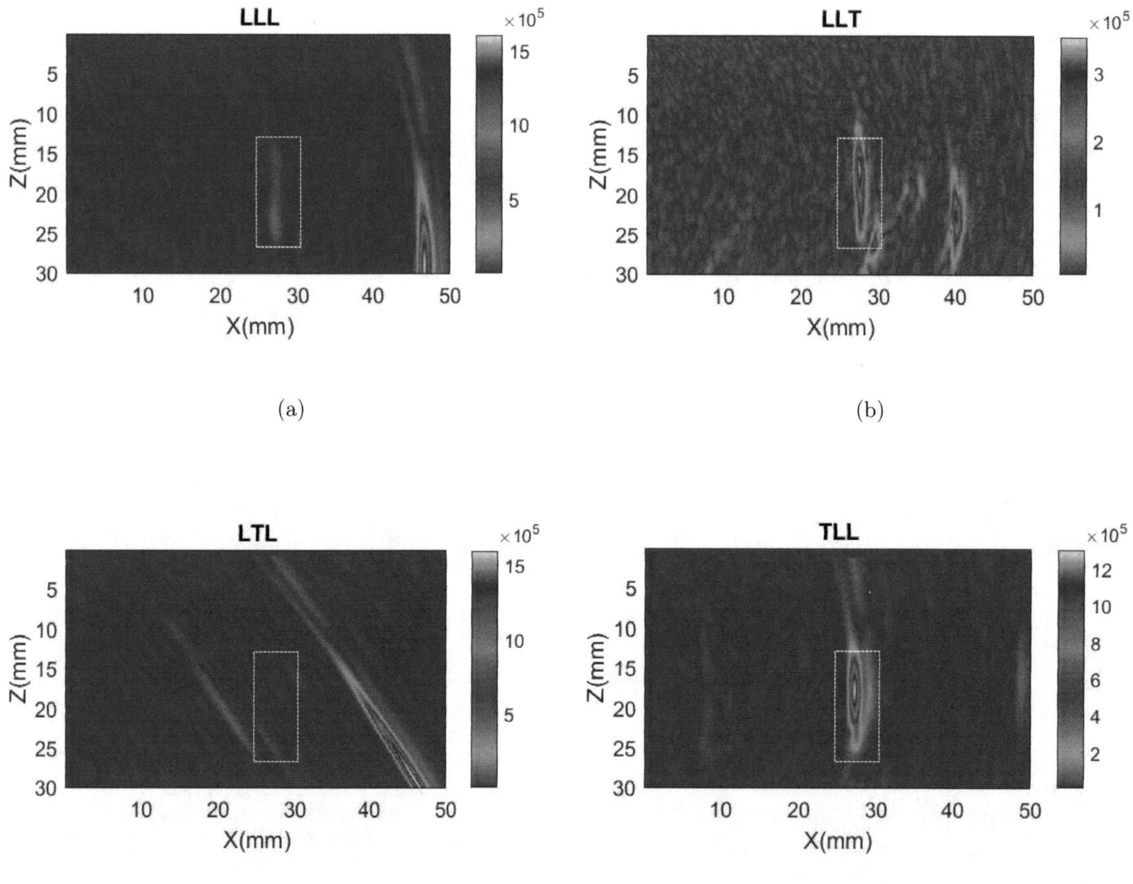

(a) (b)

(c) (d)

IOP Conf. Series: Journal of Physics: Conf. Series **1017** (2018) 012005 doi:10.1088/1742-6596/1017/1/012005

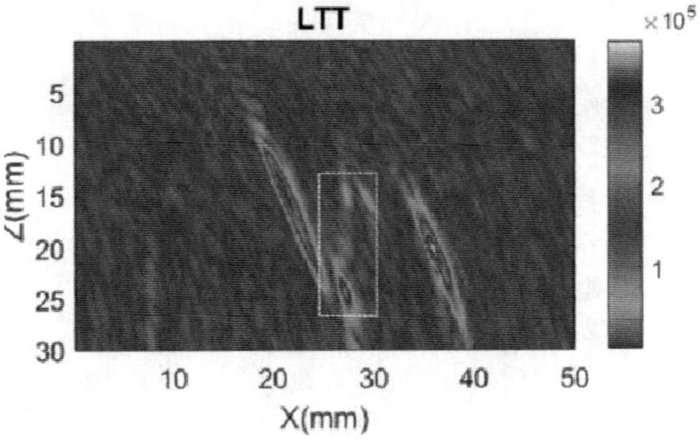

(e)

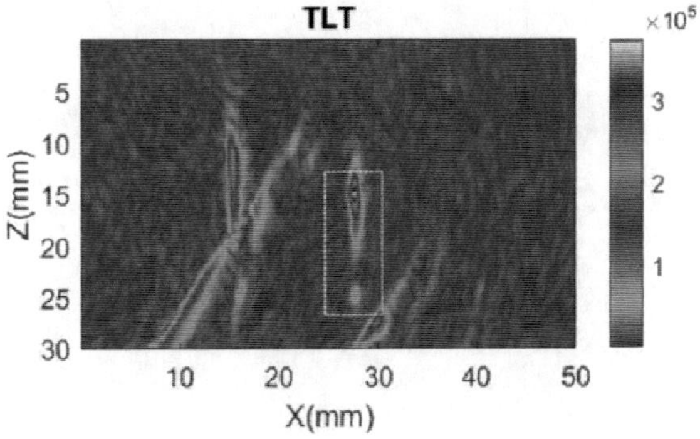

(f)

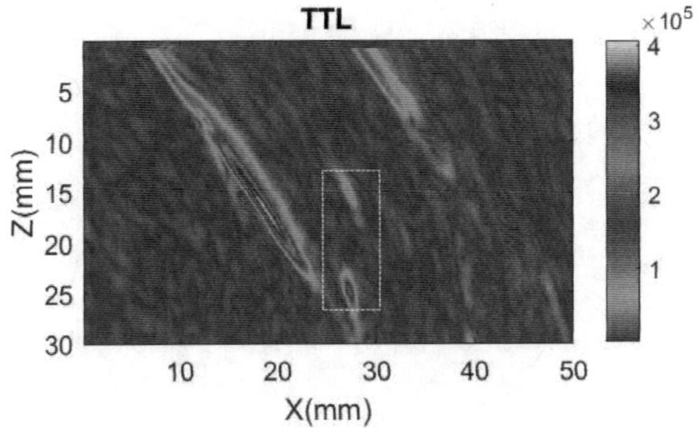

(g)

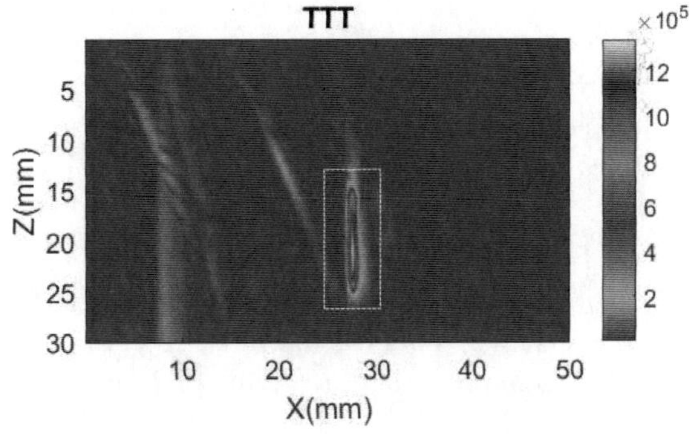

(h)

Figure 2: TFM imaging for corner echo modes

It shows that the reconstruction of the defect depends upon the mode of reconstruction, as in effect expected. In this example, 3 modes (TTT, TLL, LLT) among the total of 8 considered enable the defect to be retrieved properly, even if this modes are not perfect and present some artifacts. The other modes bring up tips diffraction, and for some, just artifacts. These latter, which look like defects, are penalizing since they indeed increase the risk of misinterpretation from the image observation.

3.2. Influence of defect position

The depth of the defect impacts the TFM reconstruction, that is, for a given mode, the reconstruction depends upon the position of the defect within the structure inspected. To illustrate this phenomenon, two cracks with same height but different positions are imaged with using the TTT mode. The first crack is at 5mm from the bottom and the second crack is at 10mm from the bottom. Each image is normalized in relation to its maximum amplitude.

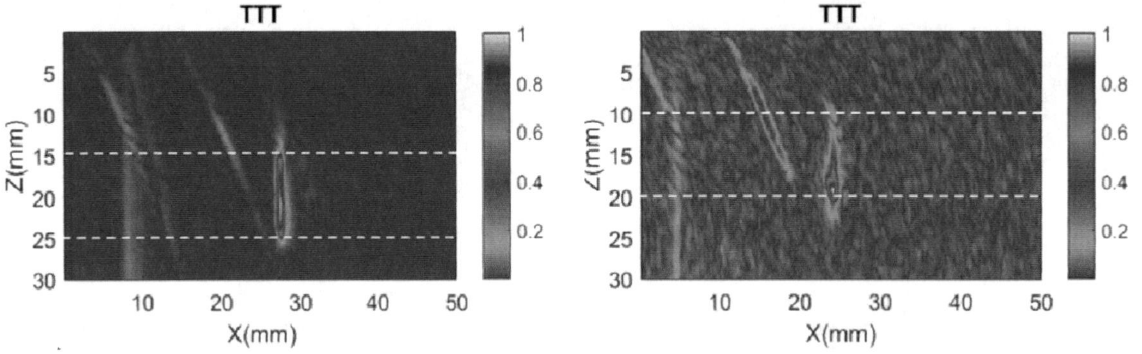

(a) (b)

Figure 3: TFM reconstruction for the crack at 5mm from the bottom (a), at 10mm (b).

The TTT mode enables the first defect to be completely retrieved, refer to Figure 3 (a), while the second one is only partially retrieved, refer to Figure 3 (b). These results show that the position of the defect is indeed crucial for its reconstruction with a given mode.

3.3. Influence of defect orientation

Besides the position, the orientation of a defect impacts the TFM imaging for a given mode. For the same configuration as in the above, two defects orientated at 10° and -10° are retrieved with two modes, the TLT and TTT ones.

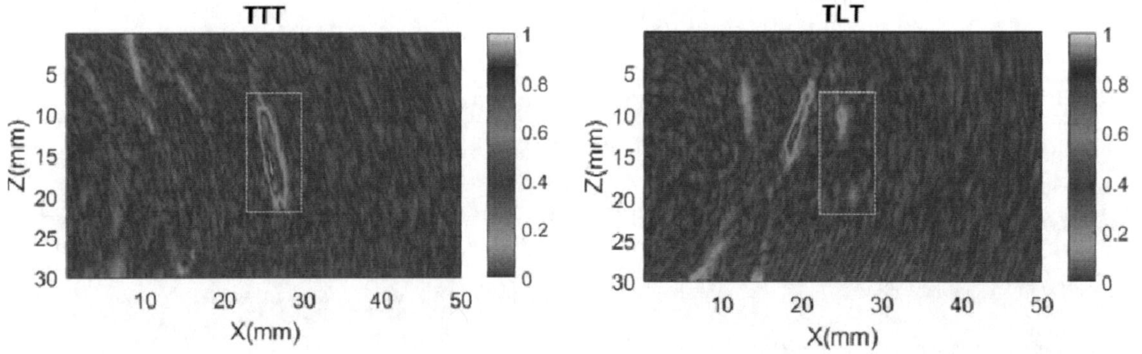

(a) (b)

The TTT mode which is appropriate for the defect orientated at 10°, refer to Figure 4(a), becomes inappropriate for the defect with the opposite orientation, refer to Figure 4(c), contrary to the TLT mode, refer to Figures 4(b) and 4(d). The choice of the pertinent mode thus becomes complicated when the orientation of the defect is unknown.

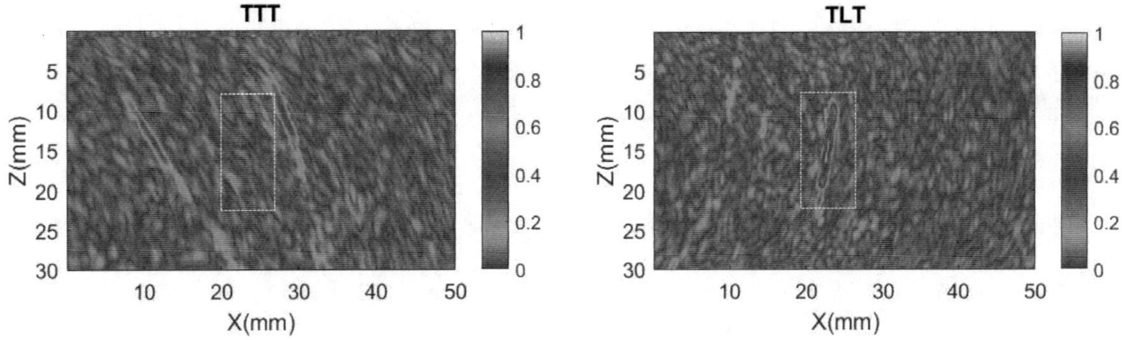

(c) (d)

Figure 4: TFM reconstruction of the crack at 10° with (a) TTT and (b) TLT modes. TFM reconstruction of the crack at -10° with (c) TTT and (d) TLT modes.

To conclude, the TFM imaging is very sensitive to the defect characteristics, and not just elsewhere. Indeed the specular reflection allowing the defect imaging is a geometrical phenomenon. So, for a couple of transmitter and receiver, the physical path is a path that obeys the Snell law on the interfaces, and also on the defect. This path is sensitive to the profile of the interfaces, the velocities of the considered waves, the orientation and the position of the defect, and even the relative position between the defect and the transducer.

Without any hypothesis on the defect, several reconstructions with several modes are needed to increase the likelihood of getting the pertinent modes, which increase greatly the number of images to analyze, which come with their own set of artefacts. The analysis mode by mode could be tedious, a method for analysing simultaneously all images is proposed next. But, before to do that the origins of artefacts are explained and a method to filter those due to the geometry echo is proposed.

4. Artefacts: origins and filtering

As shown in the preceding section, artifacts may appear in TFM imaging, being emphasized that we call artifacts all coherent indications present in the image but not corresponding to an actual defect. These artifacts, due to the coexistence of several modes in the same received signal, are derived from the defect echo or the geometry echo. In some cases they look like a defect and improve the risk of misinterpretation, and in others their amplitude is so strong that they can hide the defect indication.

4.1. Non-uniqueness of time of flight in the same mode

As described, the times of flight are not unique in TFM imaging. Indeed, in the same mode, we have isochronous curves which are ellipses of which all points are associated to the same amplitude. If a defect is present at a point, the summation of amplitudes of all couples is coherent and the TFM amplitude increases. Otherwise, it vanishes.

But in some configurations, where the elements of the phased array are not 'enough' divergent, or if the defect length is very large vs. the transducer aperture, the interferences are not sufficient to completely cancel the amplitude outside the defect location, so the quality of images is deteriorated. The removal of these artifacts involve the optimization of the inspection and is not discussed in this statement.

4.2. Non-uniqueness of time of flight in different modes
It is also possible to have the same time of flight and so the same contribution for two different modes. In this case, an indication at a given location, for a given mode, will be positioned at another location, for another mode. When the artifacts is due to the defect, we observe some echoes looking like a defect, seen in the section 2.1, and when it is due a specular geometry echo, we can have in some cases, especially in corner echo modes, an artefact with a very high amplitude. To filter it, we apply one method [8] based on the notion of specular path at the focusing point. Its principle is to attenuate all path without specular meaning by introducing a modification in the TFM algorithm. This method has been applied on experimental data. Two images before and after filtering are presented for LLL mode (cf. Figure 5).

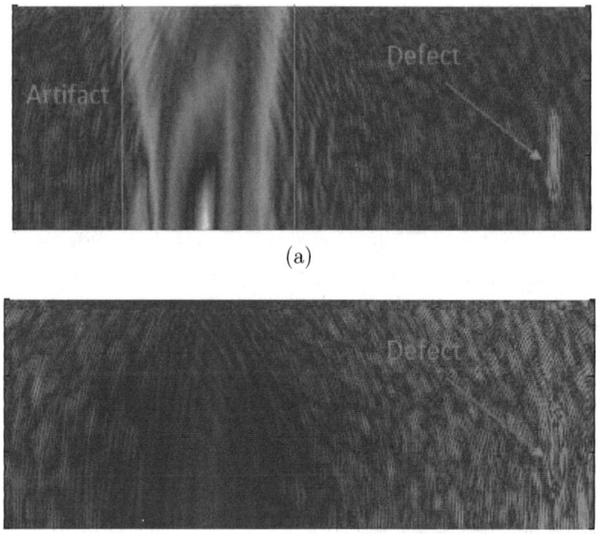

(a)

(b)

Figure 5: TFM reconstruction for LLL mode (a) without geometry artifacts filtering, (b)with geometry artifacts filtering

The artefact is completely removed by this method, refer to Figure 5(b), noticing however that the artifacts due to the defect echo are not removed by this filtering. In addition, let us specify that this algorithm works well if and only if, the geometry echo is separable from the defect echo.

Once the images filtered from geometry artifacts, several combinations of modes are possible to analyze more easily images. Combinations must be done as to do not lose information about the defect, and wherever possible, to reduce artifacts.

5. Fusion of modes by summation
It has been shown in the second section that multi-mode TFM imaging yields a large number of images which can be tedious to analyze, and in addition the pertinent information can come from several modes. So fusion of modes is a mean to simultaneously exploit all information and it can be carried out so as to filter or at least to reduce artifacts due the defect (observed in Figure 2). The fusion is performed for all modes having the same type of paths, which means the

4 direct modes between them, the 8 corner echo modes between them and the 16 indirect modes between them. So the number of images goes from 28 to 3, greatly facilitating the analysis. In this contribution the examples presented are for corner echo modes.

The summation of modes is a simple mean to not lose any information by exploiting the complementarity of modes. The final amplitude I_{sum} is such that

$$I_{sum}(P) = \sum_{m=1}^{M} |I^m(P)| \tag{2}$$

where M is the number of modes to take into account. The summation of the absolute value of images is explained by the fact that the modes have not the same phase, so they can cancel themselves on the defect by summation; however the summation of the absolute values deteriorates the SNR since the noise for all modes is added up constructively.

To illustrate the said method, the 8 images of Figure 2 are added up according to the relation (2). The defect is correctly retrieved, refer to Figure 6, and there is no loss of information.

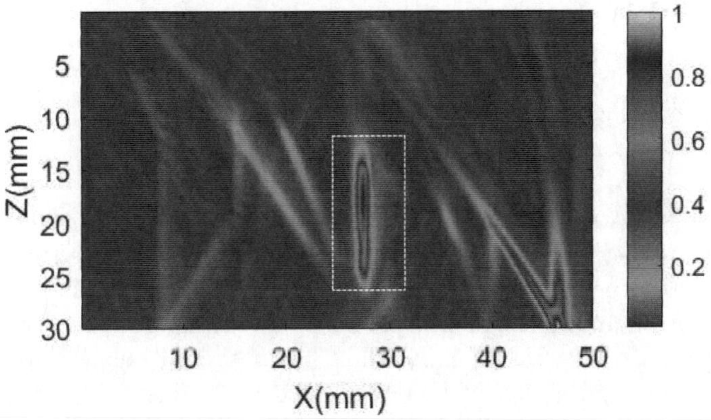

Figure 6: Fusion by summation of the 8 corner echo modes of Figure 4.

The defect is correctly retrieved, refer to Figure 6, and there is no loss of information. However, the summation brings also artifacts of all modes that are intervening in the fusion. These artifacts can be reduced if a preselection of modes is made before the summation. Indeed in this example, the most energetic artifacts come from the modes LLL and LTL (Figure 4 (a and c)), where these modes have a low refraction coefficient on the interfaces for the used transducer which mainly transmits transverse waves at 45°.

Simply adding modes with transverse waves at refraction, which are TLL, TLT, TTL and TTT, the most energetic, we observe, refer to Figure 7, that the artifacts are much reduced and the defect is properly reconstructed. In addition, the reduction of the number of modes allows to improve the SNR.

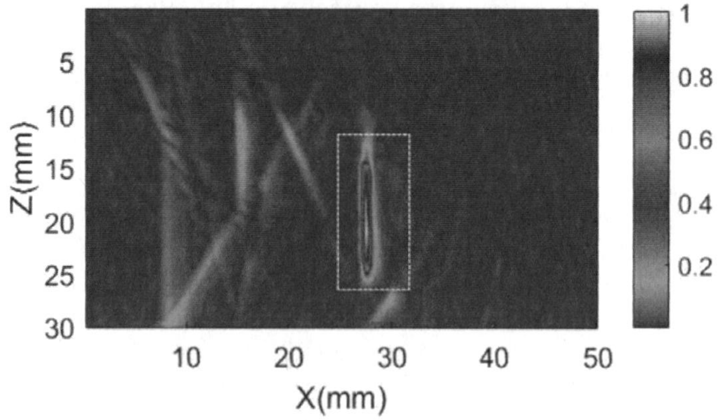

Figure 7: Fusion by summation of modes TLL, TLT, TTL and TTT.

However, preselection of modes does not overcome the artifacts fully. Fusion by summation should be preceded by filtering of artifacts due to the defects.

6. Conclusion

In this contribution, we illustrate some limits of TFM imaging, particularly its sensitivity to the defect characteristics and the reconstruction artifacts. The proposed method of artifacts filtering then applied onto experimental data shows a good capability to filter geometry artifacts. The method of fusion by summation which is proposed for exploiting simultaneously all modes presents the advantage not to lose defect indication, but it should be accompanied by methods of artifacts filtering. The preselection of modes showed a considerable reduction of artifacts, however, the energetic criterion is not sufficient to predict pertinent modes. Indeed, the specular echo also is a geometric phenomenon, the hypothesis of the knowledge of the defect orientation could permit a prediction of pertinent modes, and thus a best preselection of modes.

7. Works in progress

In the next work, a method for predicting relevant modes will be developed. This method allows to reduce considerably the artifacts and the modes to take into account for the defect reconstruction.

The method is based in the research of specular paths on the defect, requiring the knowledge of the defect orientation. It allows to predict a zone, called coverage area, in the defined region of interest, where a defect can be reconstructed. Once the coverage area is determined for a several modes, modes will be selectionned based on some criteria (maximal amplitude, maximal coverage, ...). In the case where several modes are necessary to reconstruct the defect, other methods of fusion will be proposed.

8. References

[1] Karaman M, Pai-Chi L and O'Donnell M 1995 Synthetic aperture imaging for small scale systems, *IEEE Trans. Ultrason., Ferroelec, Freq. Control* **42** 429-42

IOP Conf. Series: Journal of Physics: Conf. Series **1017** (2018) 012005 doi:10.1088/1742-6596/1017/1/012005

[2] Holmes C, Drinkwater B W and Wilcox P. D 2005 Post-processing of the full matrix of ultrasonic transmit-receive array data for non-destructive evaluation, *NDT & E Intern.* **38** 701-11

[3] Bannouf S, Robert S, Casula O and Prada C 2013 Data set reduction for ultrasonic TFM imaging using the effective aperture approach and virtual sources *J. Phys.: Conf. Series* **457** 012007

[4] Le Jeune L, Robert S, Lopez Villaverde E and Prada C 2016 Plane wave imaging for ultrasonic non-destructive testing: Generalization to multimodal imaging, *Ultrason.* **64** 128-38

[5] Zhang J, Drinkwater B W, Wilcox P.D and Hunter A. J 2010, Defect detection using ultrasonic arrays: The multi-mode total focusing method, *NDT & E Intern.* **43** 123-33

[6] Sutcliffe M, Weston M, Dutton B, Charlton P and Donne K 2012 Real-time full matrix capture for ultrasonic non-destructive testing with acceleration of post-processing through graphic hardware, *NDT & E Intern.* **51** 16-23

[7] Felice M V, Velichko A and P D. Wilcox 2014 Accurate depth measurement of small surface-breaking cracks using an ultrasonic array post-processing technique, *NDT & E Intern.* **68** 105-12

[8] Iakovleva E, Chatillon S, Bredif P and Mahaut S 2014 Multi-mode TFM imaging with artifacts filtering using CIVA UT forwards models, *AIP Conf. Proc.* **1581** 72

IOP Conf. Series: Journal of Physics: Conf. Series **1017** (2018) 012006 doi:10.1088/1742-6596/1017/1/012006

A Gaussian beam method for ultrasonic non-destructive evaluation modeling

O Jacquet[1], N Leymarie[1] and D Cassereau[2]
[1] CEA, LIST, FRA-91191 Gif-sur-Yvette, France.
[2] CNRS UMR 7371, LIB, FRA-75006 Paris, France.

E-mail: `olivier.jacquet@cea.fr`

Abstract. The propagation of high-frequency ultrasonic body waves can be efficiently estimated with a semi-analytic Dynamic Ray Tracing approach using paraxial approximation. Although this asymptotic field estimation avoids the computational cost of numerical methods, it may encounter several limitations in reproducing identified highly interferential features. Nevertheless, some can be managed by allowing paraxial quantities to be complex-valued. This gives rise to localized solutions, known as paraxial Gaussian beams. Whereas their propagation and transmission/reflection laws are well-defined, the fact remains that the adopted complexification introduces additional initial conditions. While their choice is usually performed according to strategies specifically tailored to limited applications, a Gabor frame method has been implemented to indiscriminately initialize a reasonable number of paraxial Gaussian beams. Since this method can be applied for an usefully wide range of ultrasonic transducers, the typical case of the time-harmonic piston radiator is investigated. Compared to the commonly used Multi-Gaussian Beam model [1], a better agreement is obtained throughout the radiated field between the results of numerical integration (or analytical on-axis solution) and the resulting Gaussian beam superposition. Sparsity of the proposed solution is also discussed.

1. Introduction

Ultrasonic Non-Destructive Evaluation (NDE) uses high-frequency acoustic and elastic waves to probe components without affecting their integrity. Since the industrial materials and structures are reaching high levels of complexity, the associated ultrasonic NDE technologies and diagnosis methods require up-to-date simulation tools. It means that an end-user solution should encompass many different realistic applications while providing essential results in reasonable time. However, the studied waves can propagate over a few hundreds of wavelengths before being monitored and a analyzed. Furthermore, there is a wide range of wave mechanisms potentially involved in the testing process: flaw scattering, mode conversion, dispersion, attenuation, beam forming. All these underlying physical processes do make any quantitative and efficient model challenging to develop.

In the CIVA software [2], the propagation of ultrasonic body waves is usually achieved semi-analytically with a ray-based model [3], which was initially implemented for piecewise media. This approach related to Dynamic Ray Tracing techniques [4] has also been developed for taking continuous description of welds into account [5]. In the context of high-frequency modeling, these asymptotic techniques naturally stand out in performing long time propagation efficiently while direct numerical methods, such as finite-difference, finite element or spectral methods are very costly in practice. In addition, it underlies a well-suited framework for tracking field contributions throughout complex media and through interactions with interfaces. Conversely, highly interferential features fields – namely caustics, shadows or postcritical incidences – are notoriously misdescribed by a ray-based formulation [6]. Fortunately, these so-called singularities can be circumvented inside such a convenient framework by generalizing to complex-valued paraxial quantities [7].

2. Context : radiating ultrasonic bulk wave planar transducer

The complex phase function that follows provides a decaying Gaussian profile away from the real ray path, giving rise to a localized solution commonly known as a paraxial Gaussian beam (GB). The ultrasonic field is then expanded into a superposition of several GBs. Such a complexification of the paraxial quantities introduces additional degrees of freedom into each GB. Their choice and their justification are the starting point of an extensive literature in different fields over the past three decades. We should reference contributions in Geophysics [8, 9], Acoustics [1], Quantum Mechanics [10] and radar engineering [11]. It appears that the various existing strategies developed to fix these GB parameters are strongly related to the nature of the problem. The method we propose shall then take some ultrasonic NDE particularities into account, leading us to formulate modeling assumptions.

Harmonic source term An ultrasonic bulk wave transducer, used as a transmitter, converts an electrical signal to acoustical energy and radiates a sound beam. Generally, the excitation signal is a pulse which is discribed by a central frequency and a limited bandwidth. Common Fourier analysis allows us to recover a time domain response by considering all the harmonic responses of interest, independently. Therefore, we only consider the source term as an harmonic piston radiator, oscillating at the angular frequency ω.

Propagation media Depending on the testing protocole, the active area on the face of the bulk wave transducer is immersed in a coupling media (immersion transducer) or in contact with an elastic solid (contact transducer). For sake of simplicity, the propagation media is assumed to be a homogeneous fluid of wave speed c and density ρ and we shall then model the radiated pressure field. Thus, we define $k = \omega/c$.

Source geometry Mono-element transducers can take many different shapes according to the NDE (or medical imaging) application to be carried out [12, 13]. Even limiting to the planar case, those are not necessarily of the axial symmetry (e.g. individual elements in ultrasonic arrays). Furthermore, their dimension can take value from less than a wavelength to several wavelengths. Then we only focus on baffled planar radiators, but nonetheless of any 2D shape S and amplitude profile.

3. Method : phase-space GB summation

The seminal idea behind the studied method is to initialize a GB set from a Gabor frame representation of the source term [14]. Once a properly tuned Gaussian window function is phase-space shifted for generating the frame representation, a paraxial approximation of the associated beam propagators gives rise to a fully defined GB set. We then obtain an explicit (but approximated) formulation of the full radiated field by summing up all the GB contributions. Before giving more details about the approach, the reader is invited Figure 1 to find out the geometry of the problem and some introduced notations.

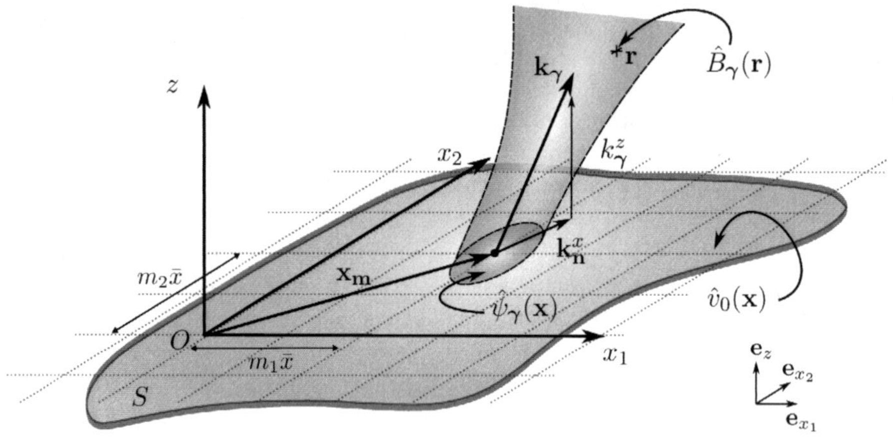

Figure 1: Gaussian Beam parametrization. A baffled planar radiator of arbitrary shape S is considered. The coordinates of any point on the radiator surface are denoted by $\mathbf{x} \equiv (x_1, x_2)$. The motion of the radiator surface acts as a piston with an arbitrary time-harmonic normal velocity $\hat{v}_0(\mathbf{x})$. A Gabor frame $\{\hat{\psi}_\gamma\}$ is introduced to decompose the source term. Each element $\hat{\psi}_\gamma$ of the frame is centered at the lattice point $(\mathbf{x_m}, \mathbf{k}^x_\mathbf{n})$ in the $(\mathbf{x}, \mathbf{k}^x)$ phase-space. The pressure field radiated into $z > 0$ by the synthesis window $\hat{\psi}_\gamma$ is evaluated asymptotically at the observation point $\mathbf{r} \equiv (\mathbf{x}, z)$ by the Gaussian beam $\hat{B}_\gamma$.

3.1. Source representation

According to the preceding assumptions, the 2D source term is known a priori and is reduced to the harmonic amplitude $\hat{v}_0(\mathbf{x})$ of the radiator normal velocity. Its Gabor frame representation is given by

$$\hat{v}_0(\mathbf{x}) = \sum_{\boldsymbol{\gamma}} \hat{a}_{\boldsymbol{\gamma}} \hat{\psi}_{\boldsymbol{\gamma}}(\mathbf{x}). \tag{1}$$

The set $\{\hat{\psi}_\gamma\}$ is defined as

$$\hat{\psi}_{\boldsymbol{\gamma}}(\mathbf{x}) = \hat{\psi}(\mathbf{x} - \mathbf{x_m}) e^{i\mathbf{k}^x_\mathbf{n} \cdot (\mathbf{x} - \mathbf{x_m})},$$

where

$$\begin{cases} \mathbf{x_m} = (m_1\bar{x}, m_2\bar{x}) \\ \mathbf{k}^x_\mathbf{n} = (n_1\bar{k}_x, n_2\bar{k}_x) \end{cases}$$

are the space and the phase shifts of the *mother window* $\hat{\psi}(\mathbf{x})$, respectively. We use the vector index $\boldsymbol{\gamma} = (m_1, m_2; n_1, n_2) = (\mathbf{m}; \mathbf{n}) \in \mathbb{Z}^2 \times \mathbb{Z}^2$ to denote the *synthesis function* $\hat{\psi}_\gamma$ unequivocally. The parameters $\bar{x}$ and $\bar{k}_x$ are the spatial and spectral displacements units respectively. For $\{\hat{\psi}_\gamma\}$ to be a frame, $\bar{x}$ and $\bar{k}_x$ must satisfy

$$\bar{x}\bar{k}_x = 2\pi\nu, \tag{2}$$

where $0 < \nu \leq 1$ is the overcompleteness factor. Once this is verified, the coefficients $\hat{a}_\gamma$ are given by

$$\hat{a}_{\boldsymbol{\gamma}} = \langle \hat{v}_0, \hat{\varphi}_{\boldsymbol{\gamma}} \rangle = \int_S \hat{v}_0(\mathbf{x}) \hat{\varphi}^*_{\boldsymbol{\gamma}}(\mathbf{x}) \mathrm{d}\mathbf{x}, \tag{3}$$

where $\hat{\varphi}^*_\gamma$ is the conjugate of the *analysis function* $\hat{\varphi}_\gamma$. The set $\{\hat{\varphi}_\gamma\}$ is the dual frame associated to the Gabor frame $\{\hat{\psi}_\gamma\}$. An explicit expression of the latter cannot be found in general, but it can be approximated for $\nu \to 0$ [14] by

$$\hat{\varphi}_\gamma(\mathbf{x}) \underset{\nu \to 0}{\approx} \left[\frac{\nu^2}{\|\hat{\psi}_\gamma\|^2} \right] \hat{\psi}_\gamma(\mathbf{x}). \tag{4}$$

Finally, we choose the *mother window* $\hat{\psi}$ to be a 2D symmetric Gaussian

$$\hat{\psi}(\mathbf{x}) = e^{-k|\mathbf{x}|^2/2z_R}, \tag{5}$$

where the significance of z_R is given below.

3.2. Paraxial Gaussian beams and field expansion

Now that we have the Gabor frame representation (1) of the source term, we want to formulate the pressure field radiated at the observation point $\mathbf{r}$. This can be represented by a plane wave integral, in particular for a given *synthesis function* $\hat{\psi}_\gamma$ we obtain the beam propagator $\hat{B}_\gamma(\mathbf{r})$ into $z > 0$,

$$\hat{B}_\gamma(\mathbf{r}) = \rho c \left(\frac{k}{2\pi} \right)^2 \int \hat{\hat{\psi}}_\gamma(\mathbf{k}^x) e^{i(\mathbf{k}^x \cdot \mathbf{x} + k^z z)} \mathrm{d}\mathbf{k}^x, \tag{6}$$

where $\hat{\hat{\psi}}_\gamma$ is the 2D spatial Fourier transform of $\hat{\psi}_\gamma$ and

$$k^z = +\sqrt{k^2 - |\mathbf{k}^x|^2}, \tag{7}$$

is the $z-$component of the wave vector $\mathbf{k} \equiv (\mathbf{k}^x, k^z)$. The integral (6) cannot be explicited directly, but it can be evaluated asymptotically by saddle point integration [15]. We then find out

$$\hat{B}_\gamma(\mathbf{r}) \approx \rho c \sqrt{\frac{\det \mathbb{M}_\gamma(z_\gamma)}{\det \mathbb{M}_\gamma(0)}} \exp \left[ikz_\gamma + \frac{ik}{2} \mathbf{x}_\gamma^T \mathbb{M}_\gamma(z_\gamma) \mathbf{x}_\gamma \right], \tag{8}$$

and we recognize the expression of a paraxial Gaussian beam formulated in ray-centered coordinates $(\mathbf{x}_\gamma, z_\gamma)$ (see [4] for example). The complex-valued curvature 2×2 matrix $\mathbb{M}_\gamma$ describes the paraxial behaviour of the radiated field around the beam axis emerging from $\mathbf{x}_m$ in the radiator plane, and supported by the wave vector $\mathbf{k}_\gamma$. The matrix $\mathbb{M}_\gamma$ only varies with z_γ and its imaginary part is positive definite for all z_γ. This complex-valued matrix explicitly depends on the ray emerging direction and z_R, introduced in (5), which is the GB collimation length (or Rayleigh length). Finally, the pressure field $\hat{p}(\mathbf{r})$ radiated by the source term is obtained by summing up all the paraxial Gaussian beams $\hat{B}_\gamma$ (8) weighted by the respective $\hat{a}_\gamma$ coefficents (3), giving

$$\hat{p}(\mathbf{r}) \approx \sum_{\gamma \in \mathbb{Z}^2 \times \mathbb{Z}^2} \hat{a}_\gamma \hat{B}_\gamma(\mathbf{r}). \tag{9}$$

3.3. Bounds and compression of the representation

As it is suggested in (9), there is an infinite number of GB contributing to the field synthesis. Fortunately, the effective GB set can be drastically shrunk in many different ways.

Spectral range Firstly, after expliciting the plane wave spectrum of $\hat{\psi}_\gamma$, we can show that $\mathbf{k}_\mathbf{n}^x$ is the single non-degenerate saddle point for the phase function of the integrand in (6). Since we only consider the propagating beams in (9) (see Figure 1) that occur only for

$$k_\gamma^z = +\sqrt{k^2 - |\mathbf{k}_\mathbf{n}^x|^2} \in \mathbb{R}_+^*, \tag{10}$$

the associated vectors index $\boldsymbol{\gamma}$ are then bounded to values of $\mathbf{n}$ that verify (10). Without deriving more explicit conditions on $\mathbf{n}$ values, $\mathcal{K}$ denotes the subset of the corresponding vectors index $\boldsymbol{\gamma}$.

Spatial range Secondly, according to (3) the amplitude of coefficient $\hat{a}_\gamma$ is non-negligible only if the associated analysis function $\hat{\varphi}_\gamma$ sufficiently overlaps the source term. Since the dual frame set $\{\hat{\varphi}_\gamma\}$ is approximated by (4), the analysis function $\hat{\varphi}_\gamma$ amplitude is a 2D Gaussian symmetric Gaussian centered in $\mathbf{x_m}$. We can then define a threshold above which the amplitude of the analysis function is thought to be significant so we can easily estimate how close to the source term the function $\hat{\varphi}_\gamma$ must be centered. Finally, given the fact that the considered radiator is baffled, the eligible vectors index γ are also bounded to such values of $\mathbf{m}$. In the same manner as above, $\mathcal{X}$ denotes the subset of the corresponding vectors index γ.

Once we take into account the spectral and the spatial ranges, (9) is reduced to

$$\hat{p}(\mathbf{r}) \approx \sum_{\gamma\in\mathcal{K}\cap\mathcal{X}} \hat{a}_\gamma \hat{B}_\gamma(\mathbf{r}). \tag{11}$$

Compression Furthermore, it is possible to take advantage of the overcompleteness of the Gabor frame representation (1) of the source term. Indeed, if $\nu = 1$, the frame $\{\hat{\psi}_\gamma\}$ is critically complete and it becomes a basis where there is a single possible representation for $\hat{v}_0$. In other words, the frame overcompleteness produces a redundancy in the representation that makes its compression possible without degrading the result of (1). In practice, this phase-space representation reveals a strong localization of the coefficients $\hat{a}_\gamma$ around clearly defined region in the $(\mathbf{x_m}, \mathbf{k_n^{\mathit{x}}})$ domain. It appears that the coefficients $\hat{a}_\gamma$ are non-negligible for phase-space points near the Lagrange Manifold [14]. Such a rather high localization is an additional argument for compression.

The compression for seeking sparse representation is a recent topic of high interest. For example, Andersson *et al* [16] apply *iterative soft thresholding* to a Gaussian wave packet representation to obtain sparse approximation of oscillatory functions. Instead of using such an iterative algorithm, we propose a compression strategy which is simply based on a single thresholding applied to coefficients $\hat{a}_\gamma$, followed by the reassignment of the discarded energy. Formally, we define a relative threshold $\epsilon < 1$ and Γ_ϵ denotes the subset of the vectors index γ corresponding to the selected coefficients $\hat{a}_\gamma$, given by

$$\Gamma_\epsilon = \left\{\gamma \in \mathcal{K}\cap\mathcal{X},\ \frac{|\hat{a}_\gamma|}{\max_{\gamma\in\mathcal{K}\cap\mathcal{X}} |\hat{a}_\gamma|} \geq \epsilon\right\}. \tag{12}$$

Since the thresholding step cuts a potentially non-negligible energy from the initial representation, the selected coefficients $\hat{a}_\gamma$ are uniformly weighted by a reassignment ratio α given by

$$\alpha = \sqrt{\frac{\sum_{\gamma\in\mathcal{K}\cap\mathcal{X}} |\hat{a}_\gamma|^2}{\sum_{\gamma\in\Gamma_\epsilon} |\hat{a}_\gamma|^2}} > 1. \tag{13}$$

The energy conservation by the weighting operation is straightforward. Finally, the phase-space bounded GB solution (11) of the radiated field $\hat{p}(\mathbf{r})$ is reduced to

$$\hat{p}(\mathbf{r}) \approx \sum_{\gamma\in\Gamma_\epsilon} \alpha\hat{a}_\gamma \hat{B}_\gamma(\mathbf{r}). \tag{14}$$

In the following section, we use (11) to evaluate the approximated radiated field, and (14) to assess the influence of the compression.

IOP Conf. Series: Journal of Physics: Conf. Series **1017** (2018) 012006 doi:10.1088/1742-6596/1017/1/012006

4. Simulation results and discussion

As an example, we consider the typical case of a 5 MHz, 6 mm diameter circular uniform piston transducer in water, giving

$$\hat{v}_0(\mathbf{x}) = \hat{v}_0, \quad |\mathbf{x}| \leq \frac{L}{2}.$$

The corresponding wavelength in the water is $\lambda = 0.3$ mm, and we can introduce the *near field distance* $N = L^2/4\lambda$ associated to a circular uniform piston radiator [17]. This quantity is well-suited for localizing the nulls and the maxima of the on-axis field. For all the following simulation results, we use an highly overcomplete frame where $\nu = 0.09$. The Rayleigh length z_R that characterizes the *mother window* $\hat{\psi}$ is chosen in such a way that the analysis functions $\hat{\varphi}_\gamma$ ensure a balanced representation with sufficient localization in both phase and space domains. Here, we arbitrary choose the width of the analysis function to be fourth the transducer diameter. In order to complete the frame definition, the spatial and spectral displacements units, $\bar{x}$ and $\bar{k}_x$ respectively, are choosen according to ν and z_R values in order to provide a full coverage of the phase-space. Thereafter, subsets $\mathcal{K}$ and $\mathcal{X}$ are easily identified. Then, noting that (3) can be expressed as windowed Fourier transforms, the coefficient set $\{\hat{a}_\gamma, \gamma \in \mathcal{K} \cap \mathcal{X}\}$ is computed quickly by performing Fast Fourier Transforms (FFTs).

Main beam region As an overview, we calculate the radiated pressure field via (11) in the main beam region that contains the transducer axis. The 2D cross-sectional result is shown Figure 2.

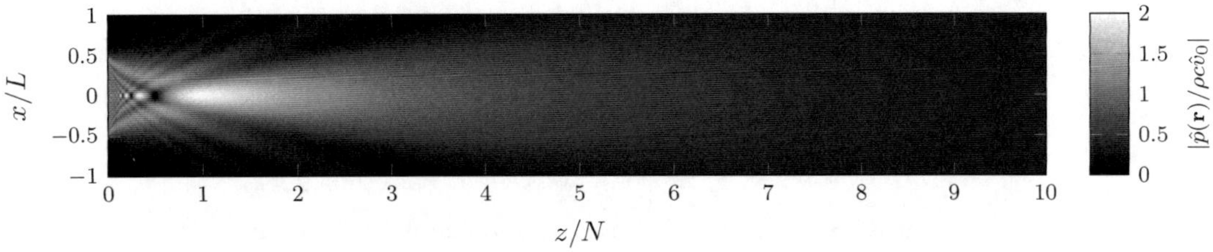

Figure 2: Cross-sectional image of the normalized magnitude of the pressure field of a 6 mm diameter circular piston transducer radiating into water at 5 MHz as calculated with the GB solution given by (11)

Qualitatively, the field structure is familiar. Even in the near field region for $z < N$, it is able to exhibit an highly interferential regime where there is a well-known alternance between nulls and maxima throughout the on-axis field, and expected side lobes throughout the cross-axis field. This calls for quantitative comparisons.

On-axis field In the case of a circular uniform piston, the on-axis pressure field is known exactly by solving the *Rayleigh-Sommerfeld integral*, giving

$$\hat{p}_{\text{ref}}(z) = \rho c \hat{v}_0 \left[e^{ikz} - e^{ik\sqrt{z^2+(L/2)^2}} \right]. \tag{15}$$

An other interesting point of quantitative comparison is provided by the Multi-Gaussiam Beam (MGB) model. Commonly used in NDE modeling, this model expands the pressure field radiated by axisymmetric transducers into a superposition of a few Gaussian beams. In this approach, the GB parameters result from an offline greedy non-linear minimization of an error function. As those parameters are wrote down in a referenced table [1], they can be re-used for convenience of implementation as they are. However, this model notoriously suffers from poor near field description and a lack of genericity and it fails in simulating steering beam fields of phased array transducers [18]. The on-axis field is computed in the near field region for $z \in [0, N/2]$ with our GB solution (11). The latter is compared to the exact solution (15) and the MGB output (see the results in Figure 3a). As expected, the MGB model does not match with the near field interferences for $z \lesssim N/4$. Conversely, our solution can keep oscillating for $z \to 0$ and reaches a better agreement than the MGB solution.

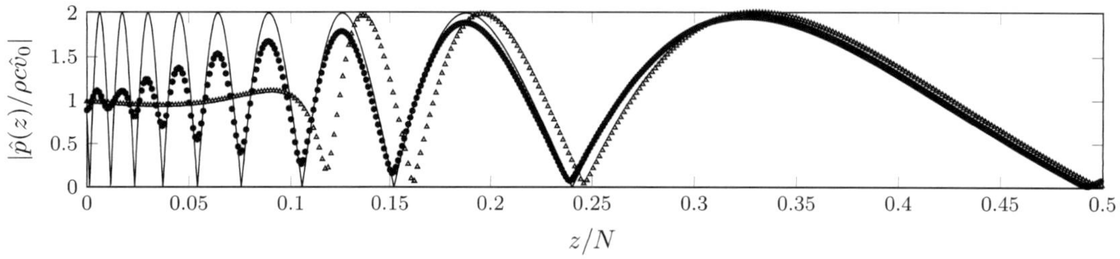

(a) Normalized magnitude of the pressure field calculated with (–) the exact solution (15), (△) the MGB model [1] and (•) the GB solution given by (11).

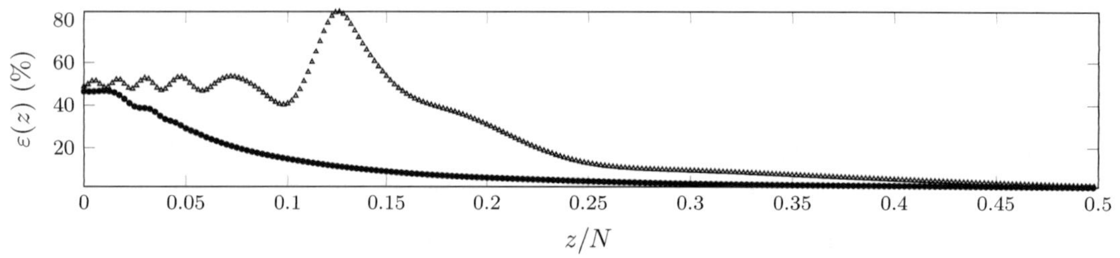

(b) Normalized absolute error level calculated for (△) the MGB model and (•) the GB solution given by (11).

Figure 3: The on-axis pressure field of a 5 MHz, 6 mm diameter circular piston transducer radiating into water is calculated in the near field region for $z \in [0, N/2]$.

Let us take a closer look at the discrepencies by calculating the normalized absolute error level ε, defined by

$$\varepsilon(\mathbf{r}) = \frac{|\hat{p}(\mathbf{r}) - \hat{p}_{\text{ref}}(\mathbf{r})|}{\max_{\mathbf{r}\in\mathbb{R}^3} |\hat{p}_{\text{ref}}(\mathbf{r})|} \tag{16}$$

where $\hat{p}$ is any approximated solution of the radiated pressure field and $\hat{p}_{\text{ref}}$ the adopted reference solution. As it is defined, the error level ε locally depicts the deviation from the reference solution in relation to the meaningful magnitude. The error levels ε related to the above comparison are shown Figure 3b. Throughout the near field region, the error curve associated to our GB solution stays below the one associated to the MGB model and decreases continuously with z. The latter aspect can be explained by the choice of discarding the evanescent modes (10) whose the contribution should get significant in the very near field region.

Cross-axis field For off-axis observation points, the harmonic response cannot be derived analytically. A natural approach consists in the numerical evaluation of the *Rayleigh-Sommerfeld integral* using a method for which the associated error is controlled [20]. Then, the chosen off-axis reference solution is the integral formulation given by Hutchins *et al* [19] and computed by a Gaussian quadrature. Radiated cross-axis pressure field is now considered at the near field distance ($z = N$) and in the very near field region ($z = N/4$), for $x \in [-L, L]$. The results are shown Figure 4 and Figure 5, respectively.

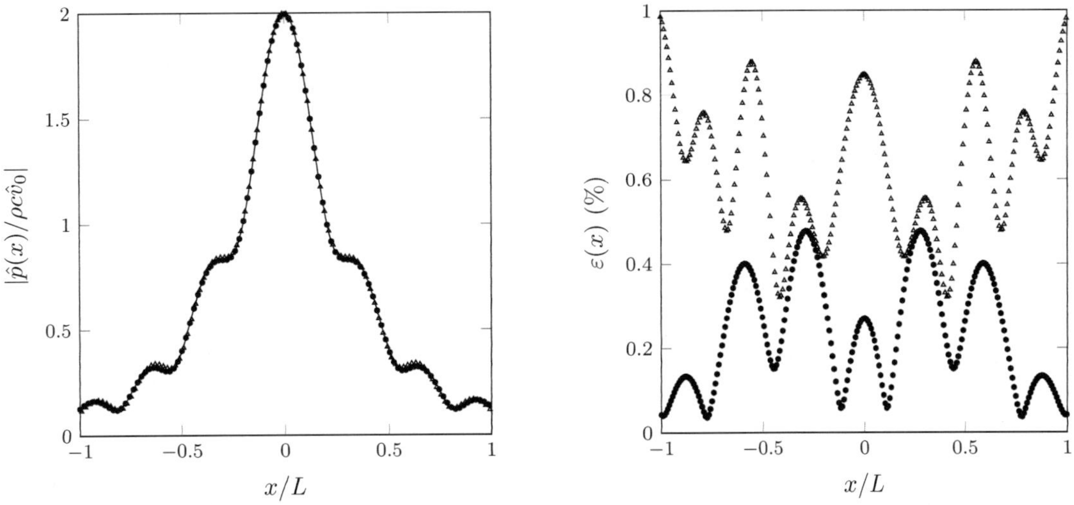

(a) Normalized magnitude of the pressure field calculated with (–) the reference solution [19], (△) the MGB model [1] and (•) the GB solution given by (11).

(b) Normalized absolute error level calculated for (△) the MGB model and (•) the GB solution given by (11).

Figure 4: The cross-axis pressure field of a 5 MHz, 6 mm diameter circular piston transducer radiating into water is calculated at the near field distance $z = N$ for $x \in [-L, L]$.

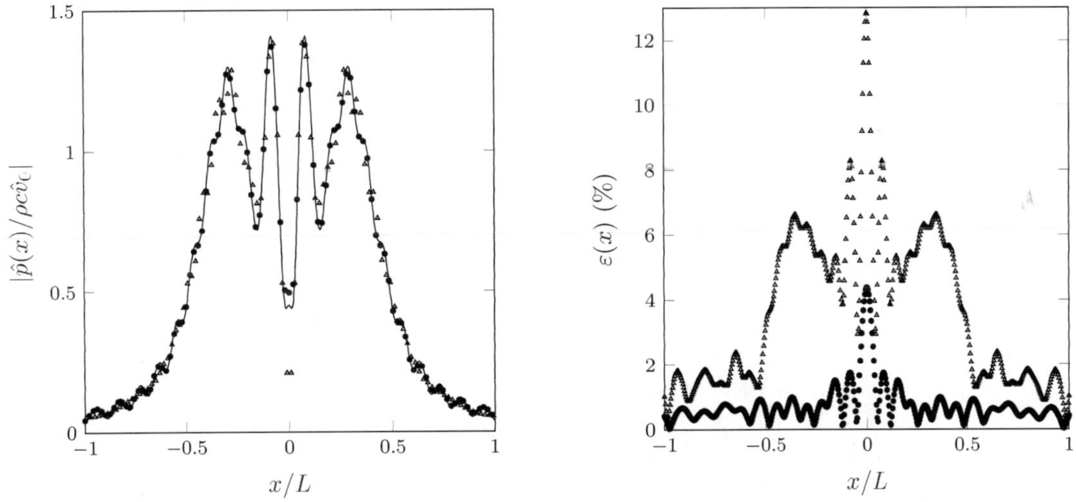

(a) Normalized magnitude of the pressure field calculated with (–) the reference solution [19], (△) the MGB model [1] and (•) the GB solution given by (11).

(b) Normalized absolute error level calculated for (△) the MGB model and (•) the GB solution given by (11).

Figure 5: The cross-axis pressure field of a 5 MHz, 6 mm diameter circular piston transducer radiating into water is calculated in the near field region at the distance $z = N/4$ for $x \in [-L, L]$.

For $z = N$, the MGB solution and our GB solution are indistinguishable from the reference (see Figure 4a), so that the main beam and side lobes are very well described. For the studied region of interest, the obtained error ε is less than 1% in both cases (see Figure 4b). Furthermore, our GB solution is even more precise than the MGB model as its associated error is strictly inferior to the one calculated with the MGB model. For $z = N/4$, the MGB solution appears to be less accurate than our GB solution. It fails in reproducing the side oscillations and it underestimates the field magnitude in the vicinity of the transducer axis (see Figure 5a). A closer look at the error levels attests those differences, where the calculated error ε for our GB solution is still reasonable (less than 5%) and substantially below the one calculated for the MGB model (see Figure 5b).

4.1. Compressed representation

This section is devoted to assess the impact of the compression introduced in (12) and (13) to the deduced GB solution in (14). This is achieved by evaluating the on-axis pressure for $z \in [0, 10N]$ with (14) for different values of the threshold coefficient $\epsilon > 0$. Then we determine the peak error level, $\max_z \varepsilon(z)$, between these evaluated fields and the uncompressed result ($\epsilon = 0$). In addition, we calculate the associated compression rate τ_ϵ that measures the part of GBs rejected following the thresholding operation. The latter is given by

$$\tau_\epsilon = 1 - \frac{|\Gamma_\epsilon|}{|\mathcal{K} \cap \mathcal{X}|}.$$

The results are listed in Table 1.

Table 1: Compression rate τ_ϵ and peak error ε due to compression for different threshold coefficient ϵ values.

ϵ	τ_ϵ	$\max_{z\in[0,10N]} \varepsilon(z)$
0	0%	0%
1.10^{-3}	71.5%	0.05%
5.10^{-3}	88.9%	0.97%
1.10^{-2}	92.2%	3.89%
2.10^{-2}	95.3%	15.9%
3.10^{-2}	96.9%	29.5%

For $\epsilon = 1.10^{-3}$, one notes that there's already a meaningful shrinkage of the GB set. It remains less than on-third of the initial number of GBs. Such a resulting compression ratio $\tau_\epsilon = 71.5\%$ for this low threshold value supports the idea that the coefficients $\hat{a}_\gamma$ are strongly localized in the $(x_\mathbf{m}, \mathbf{k}^x_\mathbf{n})$ domain, as mentioned above. Yet one observes that the consecutive error $\varepsilon(z)$ calculated throughout the z-axis does not exceed a marginal level of 0.05%. Further compression given by $\epsilon = 1.10^{-2}$ leads to 92.2% of the initial GBs to be rejected while the generated error ε is still under 4%. Reaching such a shrinkage while the final solution is weakly affected would not have been possible without the reassignment (13) of the discarded energy.

5. Conclusion and perspectives

The purpose of this paper was to derive a rather generic approach to expand an ultrasonic field into a superposition of several paraxial Gaussian beams for NDE modeling applications. Since it is well-known that any strategy devoted to initialize Gaussian beam depends heavily on the typology of the source term and the nature of the propagation media, different modeling hypothesis have been formulated in keeping with the ultrasonic NDE requirements. Without being too restrictive, we have considered the field radiated into a fluid by baffled planar transducers of any 2D shape, vibrating as an harmonic piston radiator. Such a framework allows us to initialize a paraxial Gaussian beam set from a Gabor frame representation of the source term. The radiated field is then obtained analytically by summing up all the Gaussian beam contributions.

This method has been tested and validated on the typical case of a circular piston transducer. We noted that our approach gives better results than the commonly used Multi-Gaussian Beam model [1], in particular in the near field region. It then provides a relevant alternative for modeling unconventional transducers of non-symmetric shapes and/or of non-uniform amplitude profile. However, it remains that evanescent modes have been discarded by necessity, leading to an underestimated field magnitude in the vicinity of the transducer active area. An other feature which has not been presented here is that for a given far field region of interest, our GB solution can be improved by choosing an higher GB collimation distance z_R. In contrast, the resulting field will be less accurate in the near field region because of a poorer space localization. Some technical considerations about frame tuning and re-expansion in reflective media can be found in [21].

Moreover, we assessed the influence of a compressed–reassigned representation of the source term to the approximated radiated field. This shed light on sparsity capabilities in setting up a modeling approach that relies on any redundant frame representation [16]. Such a feature is encouraging for potentially more complex media since it greatly reduces the number of Gaussian beams to be tracked.

Acknowledgments

O Jacquet would like to thank A Aubertin for his technical contribution and for his interest in this work.

References

[1] Wen J J and Breazeale M A 1988 A diffraction beam field expressed as the superposition of Gaussian beams *J. Acoust. Soc. Am.* **83** 1752–56
[2] Calmon P, Mahaut S, Chatillon S and Raillon R 2006 CIVA: an expertise platform for simulation and processing NDT data *Ultrasonics* **44** e975–979
[3] Gengembre N and Lhémery A 2000 Pencil method in elastodynamics: application to ultrasonic field computation *Ultrasonics* **38** 495–9
[4] Červený V 2001 *Seismic Ray Theory* (Cambridge: Cambridge University Press)
[5] Gardahaut A, Jezzine K, Cassereau D and Leymarie N 2014 Simulation of ultrasonic wave propagation in welds using ray-based methods *J. Phys.: Conf. Series* **498** doi:10.1088/1742-6596/498/1/012008
[6] Runborg O 2007 Mathematical models and numerical methods for high frequency waves *Commun. Comput. Phys.* **2**, 827–80.
[7] Červený V, Popov M M and Pšenčík I 1982 Computation of wave fields in inhomogeneous media – Gaussian beam approach *Geophys. J. Int.* **70** 109–28
[8] Klimeš L 1984 Expansion of a high-frequency time-harmonic wavefield given on an initial surface into Gaussian beams *Geophys. J. Int.* **79** 105–18
[9] Hill N R 2001 Prestack Gaussian-beam depth migration *Geophysics* **66** 1240–50
[10] Leung S and Qian J 2009 Eulerian Gaussian beams for Schrödinger equations in the semi-classical regime *J. Comput. Phys.* **228** 2951–77
[11] Chabory A, Sokoloff J and Bolioli S 2008 Physically based expansion on conformal Gaussian beams for the radiation of curved aperture in dimension 2 *IET Microwaves Antennas Propag.* **2** 152–7
[12] Hunt J W, Arditi M and Foster F S 1983 Ultrasound transducers for pulse-echo medical imaging *IEEE Trans. Biomed. Eng.* **BME-30** 453–81
[13] Kino G S 1987 *Acoustic Waves: Devices, Imaging, and Analog Signal Processing* (Upper Saddle River: Prentice Hall)
[14] Shlivinski A, Heyman E, Boag A and Letrou C 2004 A phase-space beam summation formulation for ultrawide-band radiation *IEEE Trans. Antennas Propag.* **52** 2042–56

[15] Heyman T 1997 Phase space beam summation: a local spectrum analysis of time dependent radiation *J. Electromagn. Waves Appl.* **11** 739–73
[16] Andersson F, Carlsson M, and Tenorio L 2012 On the representation of functions with Gaussian wave packets. *J. of Fourier Anal. and Appl.* **18** 146–81
[17] Schmerr L W 2016 *Fundamentals of Ultrasonic Nondestructive Evaluation* (Berlin: Springer)
[18] Huang R, Schmerr L W and Sedov A 2008 Modeling the radiation of ultrasonic phased-array transducers with Gaussian beams *IEEE Trans. Ultrason., Ferroelect., Freq. Control.* **55**(12)
[19] Hutchins D A, Mair H D, Puhach P A and Osei A J 1986 Continuous-wave pressure fields of ultrasonic transducers *J. Acoust. Soc. Am.* **80** 1–12
[20] McGough R J, Samulski T V and Kelly J F 2004 An efficient grid sectoring method for calculations of the near-field pressure generated by a circular piston *J. Acoust. Soc. Am.* **115** 1942–54
[21] Ghannoum I, Letrou C and Beauquet G 2010 Frame based Gaussian beam bouncing *2010 URSI Int. Symp. on Electromagnetic Theory (EMTS) (Berlin)* 68–71

IOP Conf. Series: Journal of Physics: Conf. Series **1017** (2018) 012007 doi:10.1088/1742-6596/1017/1/012007

A hybrid method combining the surface integral equation method and ray tracing for the numerical simulation of high frequency diffraction involved in ultrasonic NDT

M Bonnet[1], **F Collino**[2], **E Demaldent**[3], **A Imperiale**[3], **L Pesudo**[1,3]

[1] POEMS, CNRS-INRIA-ENSTA UMR 7231,
ENSTA Paristech, 828 boulevard des Marchaux, 91120 Palaiseau, France
[2] 39 rue de Montreuil 75011 Paris
[3] CEA, institut LIST, Centre de Saclay, Bat 565 Digiteo Labs, 91191 Gif-sur-Yvette, France

E-mail: `laure.pesudo@ensta-paristech.fr`

Abstract. Ultrasonic Non-Destructive Testing (US NDT) has become widely used in various fields of applications to probe media. Exploiting the surface measurements of the ultrasonic incident waves echoes after their propagation through the medium, it allows to detect potential defects (cracks and inhomogeneities) and characterize the medium. The understanding and interpretation of those experimental measurements is performed with the help of numerical modeling and simulations. However, classical numerical methods can become computationally very expensive for the simulation of wave propagation in the high frequency regime. On the other hand, asymptotic techniques are better suited to model high frequency scattering over large distances but nevertheless do not allow accurate simulation of complex diffraction phenomena. Thus, neither numerical nor asymptotic methods can individually solve high frequency diffraction problems in large media, as those involved in UNDT controls, both quickly and accurately, but their advantages and limitations are complementary. Here we propose a hybrid strategy coupling the surface integral equation method and the ray tracing method to simulate high frequency diffraction under speed and accuracy constraints. This strategy is general and applicable to simulate diffraction phenomena in acoustic or elastodynamic media. We provide its implementation and investigate its performances for the 2D acoustic diffraction problem. The main features of this hybrid method are described and results of 2D computational experiments discussed.

1. Motivations

Ultrasonic Non-Destructive Testing (US NDT) aims at probing continuous media by propagating high frequency waves. During an Ultrasonic Non-Destructive test, a wave is generated outside the material and propagates through the medium, bouncing back against interfaces and diffracting from edges or small defects (inclusions, cracks...). Its propagation response is detected by receivers placed at the surface of the medium and analyzed to obtain useful information on the structure and integrity of the probed material. Numerical simulation has become an important tool to understand and interpret those measurements, allowing to detect and characterize surface breaking or internal defects. It has also become of great interest to help designing and optimizing inspection protocols. In a nutshell, the development of high frequency cost-effective simulation tools well-adapted for industrial use progressively made it possible for US NDT to be assisted by numerical simulations. The computational strategy of these tools is of major importance and should both allow fast computations and good approximations of complex diffraction phenomena.

In the second half of the 20th century, asymptotic methods were specifically designed to deal with wave scattering problems at high frequency regimes. They allow fast non-dispersive computations of the wave scattering over large scales and give quantitative predictions of its diffraction in a large range of configurations. However, they fail in predicting accurately the complex phenomena occuring at the wavelength scale, e.g. diffraction by small defects, or interactions between multiple defects or a defect and an interface. Besides, they lack generality as those methods do not take into account the physical phenomena arising at the different scales, such as scattering over large distances, specular reflection, transmission across an interface, diffraction by a corner. All those phenomena require additional scale-specific asymptotic modeling.

On the other hand, classical numerical methods, such as the finite difference (FD) method, the finite element method (FEM) or the boundary element method (BEM) allow to compute a very accurate numerical approximation of the complex multiscale diffraction problem involved in the NDT configurations and modeled by a partial differential equation. They nevertheless entail excessive time and memory costs at high frequencies.

Consequently, neither family of methods can individually solve high frequency diffraction problems in large media both fast and accurately. However, their combination into a hybrid strategy leads to a good compromise between accuracy and computational cost for such critical configurations. In this work we propose a new hybrid strategy that combines the surface integral equation method [3] and the ray tracing method [6]. This hybrid method relies on a two-scale model of a diffracting defect, which allows to separate the treatement of the propagation and diffraction phenomena. The propagation effects are computed using the ray tracing asymptotic method, a Lagrangian formulation of geometrical optics based on a WKB ansatz for the solution of the wave equation [6]. The diffraction phenomena are predicted accurately by the BEM.

This paper is devoted to the formulation and computational demonstration of our multi-scale hybrid strategy. Section 2 introduces the model problem, for which the hybrid solution strategy is formulated in section 3. Then an accelerated version of this strategy is outlined in section 4 in view of NDT applications. Finally, numerical results that demonstrate implementation of the hybrid method for a 2D acoustic model problem are presented in section 5.

2. Model problem

We consider a model problem of 2D time-harmonic acoustic scattering of an incident wave u^{i}, propagating in a large bounded medium Ω, by a sound-soft diffracting obstacle $\Omega_0 \subset \Omega$ with the boundary Γ, as illustrated in Figure 1. The model problem is described by the Dirichlet problem for the 2D Helmholtz equation,

$$\begin{cases} \Delta u + k^2 u = 0, & \text{in } \Omega, \\ u = -u^{\mathrm{i}}, & \text{on } \Gamma, \end{cases} \tag{1}$$

where u is the unknown diffracted field. The incident field u^{i} is created by an excitation in the defect-free domain Ω. Additionally to (1), u satisfies some (unspecified) homogeneous boundary conditions on $\partial\Omega$, of the same type as the ones verified by u^{i}. We denote by x_s and x_r respectively the positions of the source S and the observation point or receiver R, and $k = \frac{2\pi f}{c}$ is the wave number associated to the incident wave emission frequency f and the medium velocity c. Some important geometrical hypotheses must now be added to describe the setting of the model

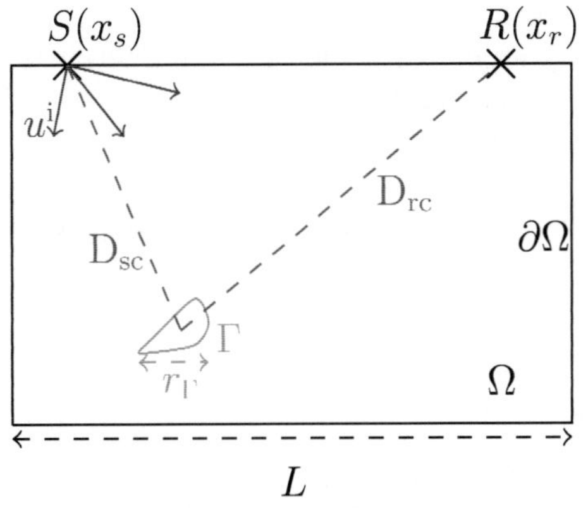

Figure 1: Diffraction configuration for the model problem.

problem (figure 1). First, the obstacle Ω_0 is supposed to be impenetrable and diffracting. Its characteristic size r_Γ is defined as

$$r_\Gamma = 2\max_{y\in\Gamma}(|y - y_c|),$$

as the diameter of the smallest disk containing Ω_0 and centered at the defect centroid y_c. It is assumed to satisfy $r_\Gamma = O(\lambda)$. Second, the propagating medium Ω is supposed to be large with respect to the wavelength ($L \gg \lambda$), isotropic and homogeneous. We assume the velocity of the wave in Ω $c = 1$. Third, both the source and the receiver are supposed to be located far enough from the defect to allow an accurate approximation of the incident and the diffracted fields by rays impinging on, or emanating from, the defect. Setting $\mathrm{D}_{\mathrm{sc}} = |x_s - y_c|$ and $\mathrm{D}_{\mathrm{rc}} = |x_r - y_c|$, this hypothesis requires satisfying $k\mathrm{D}_{\mathrm{sc}} \gg 1$ and $k\mathrm{D}_{\mathrm{rc}} \gg 1$. Finally, we suppose the defect to be located far enough from the medium boundary $\partial\Omega$ so that the interactions of the incident and diffracted fields with $\partial\Omega$ can be viewed and computed as reflections in the sense of geometrical optics. Under this framework, the contributions of the reflections of the incident and diffracted rays on $\partial\Omega$ are neglected on the defect. We approximate the incident field at each point of Γ $(x_j)_{j\in[1:N_d]}$ as a ray characterized by its amplitude A_j^s and phase ϕ_j^s:

$$u_{\mathrm{inc}}(x_j) = A_j^s e^{i\omega\phi_j^s}, \tag{2}$$

and the diffracted field at x_r as a sum of all the rays $(B_j^r, \phi_j^r)_{j\in[1:N_d]}$, emitted towards x_r from points of Γ:

$$u(x_r) = \sum_{j=1}^{N_d} B_j^r e^{i\omega\phi_j^r}. \tag{3}$$

For more accuracy during a Non-Destructive inspection, the sources and receivers are in practice sequentially moved around the probed medium. Consequently, realistic NDT configurations demand to compute the diffracted field at multiple positions of the receiver for multiple source positions. However, at ultrasonic regimes, an accurate representation of those fields requires a spatial discretization of Γ whose meshsize scales with $1/f$, making representations (2) and (3) computationally very expensive. The hybrid strategy proposed in this work relies on a two-scale model of the obstacle Ω_0 that allows to approximate the incident and diffracted field on Γ using a small number of rays.

IOP Conf. Series: Journal of Physics: Conf. Series **1017** (2018) 012007 doi:10.1088/1742-6596/1017/1/012007

3. The two-scale solution approach

The high frequency diffraction model problem is a multi-scale problem. We can identify at least two scales with respect to the wavelength on which the physics of the wave propagation is different: the propagating medium scale $L = O(100\lambda)$, corresponding to the high frequency wave propagation phenomena, and the obstacle scale, $O(\lambda)$, at which the interaction of the incident wave with the obstacle Γ generates diffraction phenomena. The hybrid strategy proposes a two-scale model of the obstacle Γ which allows to dissociate computations of propagation and diffraction phenomena. The solution of the model problem by the hybrid strategy is thus decomposed into three independent stages:

- the propagation of the incident wave u^{i} from the source S to the obstacle,
- the diffraction by the obstacle,
- the propagation of the scattered field from the obstacle to the receiver R.

Each stage is solved separately by a scale-specific method. The high frequency propagation phenomena being well modeled by Geometrical Optics laws, the two propagation problems are solved by the asymptotic ray tracing method. For those stages, set on the propagation (large) scale, the obstacle, is modeled by means of a small number P of well-chosen points close to Γ which allow to approximate the incident and diffracted fields on Γ as the plane wave approximations of the rays received at, or emitted by, those points. Diffraction coefficients, which carry information about the obstacle geometry and characteristics, are associated to these points, called centers and denoted by $(y_c^p)_{p\in[1:P]}$. Solving the propagation problems thus comes down to shooting rays from the source to each y_c^p and from the latter to the observation point. We hence expect a fast computation of the propagation phenomena. The validity of this model will be discussed in more detail in sections 3.1 and 3.2.

The diffraction phenomena are computed by the integral equation method in order to accurately describe the physics of the scattering problem at the obstacle scale. For this stage, the obstacle is described by its exact geometry. Still, in the model problem setting, the interactions of diffracted waves with the medium boundary $\partial\Omega$ are modeled as high frequency propagation phenomena, consequently the scattering by Γ is computed as the outgoing solution of the scattering problem in an unbounded medium:

$$\begin{cases} \Delta u + k^2 u = 0, & \text{in } \mathbb{R}^2 \setminus \bar{\Omega}_0, \\ u = -u^{\mathrm{i}}, & \text{on } \Gamma, \\ \lim\limits_{|x|\to\infty} \sqrt{|x|}\left(\dfrac{\partial u}{\partial |x|} - ik\,u\right) = 0. & \end{cases} \tag{4}$$

Finally, the results of the three stages are recombined to evaluate the diffracted field at the receiver.

Before getting into more details on the description of the two-scale solution approach outlined above, we briefly describe the methods chosen to compute the propagation and diffraction stages.

The ray tracing method. The ray tracing method is a Lagrangian formulation of geometrical optics based on the representation of high frequency waves v by a WKB ansatz [6] having the form

$$v(x) = A(x)e^{i\omega\phi(x)}.$$

This ansatz allows to characterize the leading-order contribution to the highly oscillatory field u as $f^{-1} \to 0$ by two frequency independent functions A and ϕ, respectively the amplitude and phase of the ray, which evolve on a much coarser grid than v. They are respectively evaluated by numerical integration of the eikonal and transport equations

$$\begin{cases} |\nabla\phi(x)| = 1, \\ 2\nabla\phi(x)\cdot\nabla A(x) + A(x)\Delta\phi(x) = 0, \end{cases} \tag{5}$$

along specific trajectories called rays which are the characteristics of the eikonal equation. In particular, the rays propagating in a homogeneous medium are straight lines.

In the hybrid solution procedure, the computation of the incident and diffracted wave fields will respectively consist in evaluating the amplitudes and phases $(A^s_p(y^p_c; x_s), \phi^s_p(y^p_c; x_s))$ of the rays propagating from the source x_s to each center y^p_c knowing the incident field value at the source, and the amplitude and phase $(A^r_p(x_r; y^p_c), \phi^r_p(x_r; y^p_c))$ of the unique ray emitted by each center, with unit amplitude and zero phase, to the observation point x_r.

The integral equation method. The integral equation method seeks the radiating solution of a well posed propagation problem in the form of a surface integral representation [2, 3]. Here, to solve the diffraction problem described by (4), we consider the equivalent surface problem described by the indirect Brackhage-Werner surface integral equation

$$\frac{\Psi(x)}{2} + \int_\Gamma \left(\frac{\partial G}{\partial\nu(y)} - ikG\right)(x,y)\Psi(y)ds(y) = -u^{\mathrm{i}}(x),\ \forall x\in\Gamma, \tag{6}$$

which is well-posed for any frequency [3]. Here G denotes the fundamental solution of the 2D Helmholtz equation in an infinite medium, i.e. $G(x,y) = \frac{\mathrm{i}}{4}H^1_0(k|x-y|)$. This integral equation is then numerically solved by the boundary element method to obtain the surface density Ψ on Γ. Applying the integral representation formula

$$u(x_r) = \int_\Gamma \left(\frac{\partial G}{\partial\nu(y)} - ikG\right)(x,y)\Psi(y)ds(y), \tag{7}$$

one then computes the outgoing solution of the diffraction problem at any point $x_r \in \Omega$.

The diffraction and propagation solution methods are independent, and rely on distinct modeling of the diffracting obstacle. To recombine their results, the hybrid strategy exploits the model problem hypotheses. If the obstacle Γ is in the far field zone of the source and receiver, depending on its size, the diffracted field at x_r can be rewritten either as a ray or as a superposition of rays with an amplitude weighted by a diffraction coefficient. This approximation and its validity conditions are now described.

3.1. The barycentric method

When the diffracting obstacle Ω_0 is small enough with respect to the wavelength and situated in the far field zone of the source and the receiver, i.e. when it fulfills the conditions

$$\begin{cases} kD_{\mathrm{sc}} \gg 1, \quad \text{and} \quad kD_{\mathrm{rc}} \gg 1 \\ kr_\Gamma = O(1), \\ \dfrac{kr^2_\Gamma}{8D_{\mathrm{sc}}} \ll 1 \quad \text{and} \quad \dfrac{kr^2_\Gamma}{8D_{\mathrm{rc}}} \ll 1, \end{cases} \tag{8}$$

the defect can be modeled at the propagation scale as the neighbourhood of a single point, its centroid y_c. This point receives a single ray from the source $(A^s(y_c; x_s), \phi^s(y_c; x_s))$ and emits a single diffracted ray $(A^r(x_r; y_c), \phi^r(x_r; y_c))$ to the receiver.

IOP Conf. Series: Journal of Physics: Conf. Series **1017** (2018) 012007 doi:10.1088/1742-6596/1017/1/012007

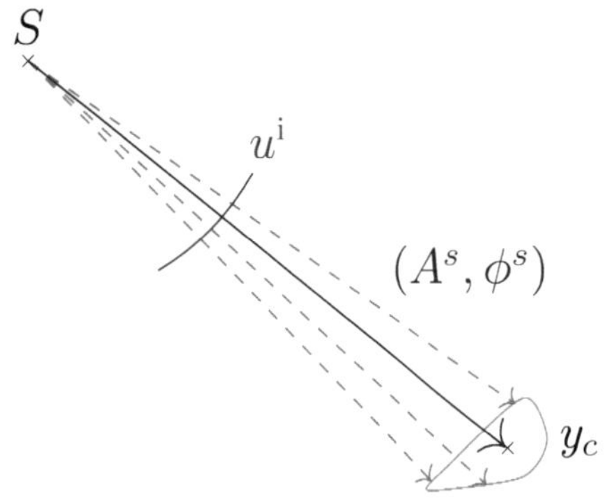

Figure 2: Ray approximation u^{i}.

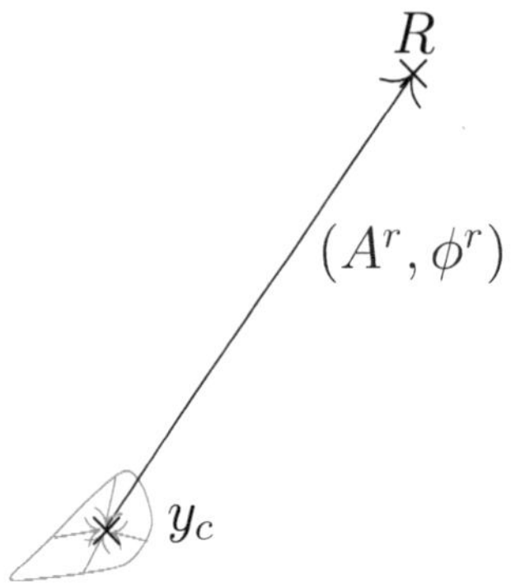

Figure 3: Ray approximation of diffracted field.

Besides, the far field hypotheses allow to link the ray approximations of the incident and diffracted fields to the solution of the diffraction problem. First, to solve the diffraction problem the incident field must be evaluated on each point of Γ from the incident ray amplitude and phase. Under conditions (8), the incident field is approximated on Γ as the linearisation $\tilde{u}^{\mathrm{i}}$ of the incident ray $(A^s(y_c; x_s), \phi^s(y_c; x_s))$:

$$\forall y \in \Gamma, \quad \tilde{u}^{\mathrm{i}}(y) := A^s(y_c; x_s)\, e^{i\omega\phi^s(y_c;x_s)}\, e^{i\omega\hat{s}(y-y_c)}, \tag{9}$$

where the unit vector $\hat{s} = \frac{y_c - x_s}{|y_c - x_s|}$ denotes the direction of propagation of the incident ray. Second, to match the hybrid strategy propagation model, the diffracted field obtained numerically by the integral representation formula (7) has to be converted into a ray. Under conditions (8), (7) can be approximated by its far field expansion:

$$u(x_r) = \frac{e^{ik|x_r - y_c|}}{\sqrt{|x_r - y_c|}}\, \mathcal{R}(\hat{r}, \hat{s})\, A^s(y_c; x_s)\, e^{i\omega\phi^s(y_c;x_s)} + o\left(\frac{1}{k||x_r - y_c|}\right), \tag{10}$$

which can be interpreted as a ray propagating from the centroid y_c to the observation point x_r, its amplitude being weighted by the complex amplitude of the incident ray at y_c and the coefficient $\mathcal{R}$ given by

$$\mathcal{R}(\hat{r}, \hat{s}) := \sqrt{\frac{-ik}{8\pi}} \int_\Gamma e^{-ik\hat{r}\cdot(y-y_c)}\, (\nu(y)\cdot\hat{r} + 1)\, \Psi(y) ds(y), \tag{11}$$

defined as the far field amplitude of (7). Here Ψ is the exact solution of the surface integral problem (6) with $u^{\mathrm{i}}(y) = e^{ik\hat{s}\cdot(y-y_c)}$.

The coefficient $\mathcal{R}$ can be interpreted as the diffraction coefficient in the direction $\hat{r}$ of the obstacle Γ illuminated by an incident plane wave of unit amplitude and direction of propagation $\hat{s}$. It only depends on the incidence and observation directions, on the frequency and on the surface Γ. Equation (9) effectively converts the diffracted field into a ray since the amplitude $A(z) = 1/\sqrt{|z|}$ and phase $\phi(z) = |z|$ functions in the order 0 WKB expansion-like function $v(|x_r - y_c|) = \frac{e^{i|x_r - y_c|\omega/c}}{\sqrt{|x_r - y_c|}}$ satisfy the eikonal and transport equations (5).

The framework of the barycentric method is efficient but may impose excessive restrictions on the obstacle size, as shown by numerical results presented in section 5. In particular the method accuracy deteriorates as the obstacle diameter increases, all other parameters being kept fixed. The hybrid strategy has thus been extended as follows to deal with bigger obstacles.

3.2. The polycentric method

This extension aims at computing the diffraction effects of a diffracting obstacle whose diameter is of order one or a few wavelengths, i.e. $r_\Gamma = O(\lambda)$. Because of its size, the obstacle is no longer in the far field zone of the source and the receiver, and will no longer be modeled as a single point at the propagation scale. The remedy consists in introducing a partition of unity of Γ, characterized by a distribution of points $y_c^p \in \Gamma$ called centers and smooth nonnegative partitioning functions $(\eta_p)_{1\leq p\leq P}$, defined on Γ and verifying

$$\sum_{p=1}^{P} \eta_p(x) = 1, \forall x \in \Gamma.$$

The supports Γ_p of the functions η_p define a covering of Γ, $\Gamma = \bigcup_{p=1}^{P} \Gamma_p$. For implementation purposes, we consider a diffracting disk and a uniform distribution of the centers y_c^p. We then define the subsets Γ_p as the intersection of Γ with the disk of center $y_c^p \in \Gamma$ and radius $r^p = |y_c^p - y_c^{p+1}|$ (with $y_c^{P+1} = y_c^1$)

$$\Gamma_p = \{y \in \Gamma \,|\, |y - y_c^p| < r^p\}.$$

The number P is chosen to guarantee that each subset is small enough to be in the far field

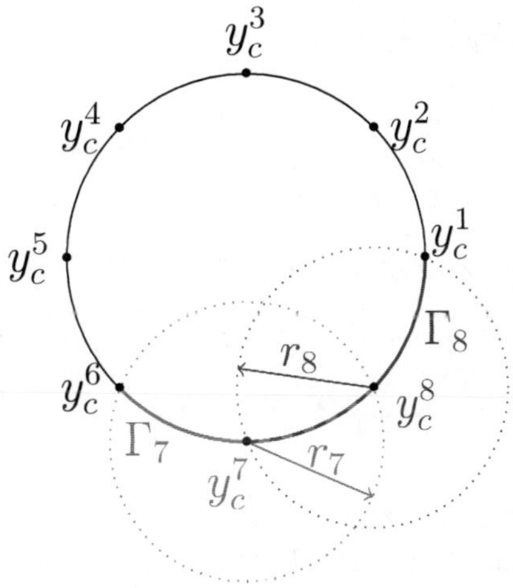

Figure 4: Covering of Γ.

zone of the source and receiver, i.e. satisfies

$$\begin{cases} kr_p = O(1), \\ \dfrac{kr_p^2}{2D_{\text{sc}}} \ll 1 \quad \text{and} \quad \dfrac{kr_p^2}{2D_{\text{rc}}} \ll 1, \end{cases} \qquad p \in [1:P]. \tag{12}$$

Under these hypotheses, the obstacle is modelled at the propagation scale as a set of P points y_c^p, each receiving a single ray propagated from the source in the direction $\hat{s}^p = \frac{y_c^p - x_s}{|y_c^p - x_s|}$ and emitting a single diffracted ray of unit amplitude and zero phase from y_c^q to the observation point x_r in the direction $\hat{r}^q = \frac{x_r - y_c^q}{|y_c^q - x_r|}$. The amplitudes and phases of those rays are computed

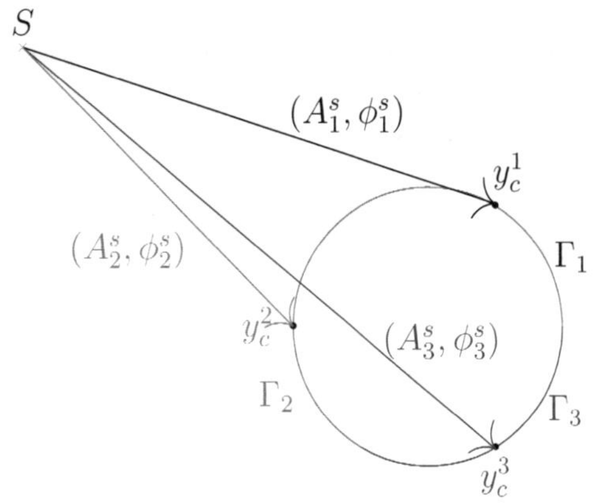

Figure 5: Ray propagation of the incident field.

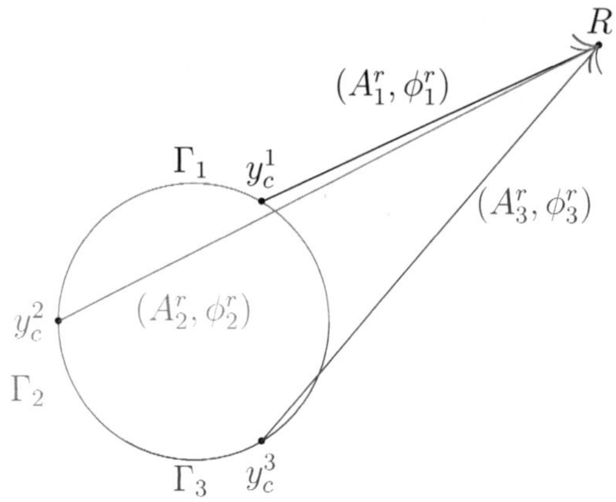

Figure 6: Ray propagation of the diffracted field.

by the ray tracing method.

As for the barycentric method, two approximations have to be introduced, one on the incident field and one on the diffracted field, to ease the recombination of the results obtained at the propagation and diffraction scales. First, to allow the solution of the diffraction problem by the integral equation method, the incident field u^{i} is approximated on each subset as the plane wave approximation $u^{\mathrm{i},p}$ of its corresponding incident ray $(A_p^s(y_c^p; x_s), \phi_p^s(y_c^p; x_s))$, i.e.:

$$\forall y \in \Gamma, \quad \tilde{u}^{\mathrm{i}}(y) = \sum_{p=1}^{P} \eta_p(y)\, \tilde{u}^{\mathrm{i},p}(y) = \sum_{p=1}^{P} \eta_p(y)\, A_p^s(y_c^p; x_s)\, e^{i\omega\phi_p^s(y_c^p;x_s)}\, e^{i\omega\hat{s}^p\cdot(y-y_c^p)}. \tag{13}$$

Second, subject to the conditions (12), the diffracted field at the receiver is converted into the superposition of P diffracted rays emitted from the centers y_c^p:

$$u(x_r) = \sum_{q=1}^{P} \frac{e^{ik|x_r - y_c^q|}}{\sqrt{|x_r - y_c^q|}} \sum_{p=1}^{P} \mathcal{R}_p^q(\hat{r}^q, \hat{s}_p)\, A_p^s(y_c^p; x_s)\, e^{i\omega\phi_p^s(y_c^p;x_s)}, \tag{14}$$

and weighted by the diffraction coefficient $\mathcal{R}_p^q$ of the qth partition when considering the contribution of the pth partition to the incident field approximation:

$$\mathcal{R}_p^q(\hat{r}^q, \hat{s}^p) = \sqrt{\frac{-ik}{8\pi}} \int_\Gamma \eta_q(y)\, e^{-ik\hat{r}^q\cdot(y-y_c^q)}\, (\nu(y)\cdot\hat{r}^q + 1)\, \Psi^p(y) ds(y), \tag{15}$$

where Ψ^p is the exact solution of the surface integral equation (6) with $u^{\mathrm{i}}(y) = e^{ik\hat{s}^p\cdot(y-y_c^p)}$.

The covering of Γ thus uncouples the role of each subset, both in terms of reception of the incident field and emission of the diffracted field allowing a muti-scale resolution. The accuracy of the evaluation of the diffracted field by this extension increases with the number of partitions (along with its computational cost).

4. Online-Offline acceleration

Considering the need for an efficient simulation of realistic NDT diffraction configurations, where multiple source and receiver positions are involved together with large numbers of pairs of incident and diffraction directions, we additionally developed an acceleration of the hybrid strategy based on an online-offline procedure. This procedure aims at reducing the computational cost of the diffraction coefficient evaluation, which represents the most expensive operation in the hybrid solution procedure. Offline, the matrix of the diffraction coefficients is evaluated at predefined incident and diffraction directions, and its truncated low rank approximation computed by a SVD [8]. Online, the polynomial interpolation of the singular functions in the direction of interest is evaluated and used to compute an approximation of the diffraction coefficients in those directions. The acceleration was found in 2D numerical tests to be effective if the number of evaluation directions exceeds about 200, and is expected to perform even better under 3D conditions.

5. Numerical results

This section presents the numerical results of the implementation of the barycentric and polycentric versions of our hybrid strategy to a 2D acoustic problem. We consider a circular defect of characteristic size r_Γ, a source point S situated at 120λ from the defect centroid y_c and an annulus of receivers around Γ bounded by circles centered at y_c and of radius 5λ, 65λ, as depicted on Figure 7. The range of distances of the receivers to the centroid are representative of the typical distances between the probe and the controlled region in realistic NDT experiments. The numerical results presented here aim at showing that the hybrid strategy proposed allows a good accuracy in the whole range of relevant distances.

5.1. The barycentric solver

We begin by testing the performances of the barycentric method on the chosen configuration for a diffracting obstacle Γ of size $r_\Gamma = 0.3\lambda$. The barycentric method computes the diffracted field with good accuracy, as shown on the map of relative error of the diffracted field with respect to the reference solution obtained on the annulus of receivers (Figure 8) by the highly accurate Nystrom method. Besides, we notice that the relative error decreases quickly with the distance of y_c to the receiver, as shown in Table 1, where the angular maximum of the relative error, denoted by $\max_\theta(\text{er})$, is given for increasing observation distances D_{rc}. The largest relative error reached at 5λ is 1.2%, indicating a resonably good precision of the barycentric method in the near field given the order of accuracy that the ray tracing method allows. If we now consider

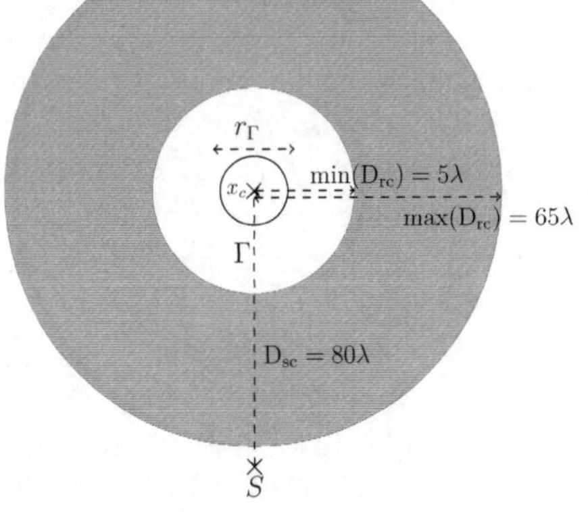

Figure 7: Numerical experiment configuration.

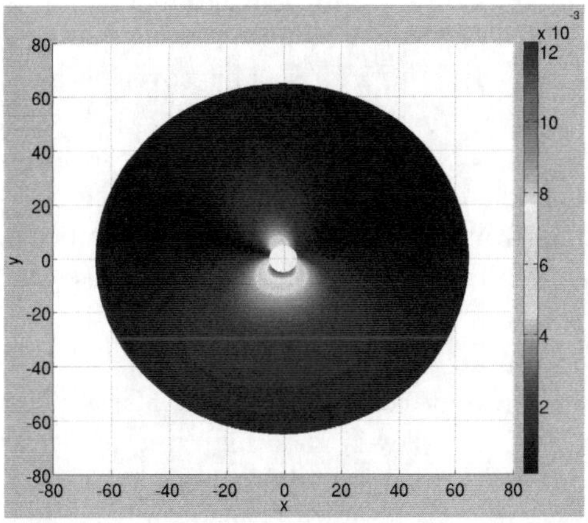

Figure 8: Map of relative error er of the diffracted field for a barycentric resolution, $r_\Gamma = 0.3\lambda$.

IOP Conf. Series: Journal of Physics: Conf. Series **1017** (2018) 012007 doi:10.1088/1742-6596/1017/1/012007

Table 1: Evolution of the relative error with respect to the distance to Γ, $r_\Gamma = 0.3\lambda$.

D_{rc}/λ	5	10	20	30	40	50	60	65
$\max_\theta(\text{er})$ ($\times 10^{-3}$)	12.52	5.99	3.02	2.02	1.51	1.21	1.01	0.93

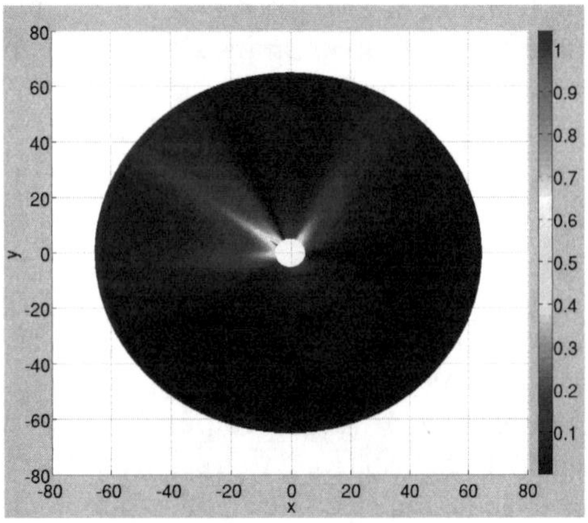

Figure 9: Map of relative error er of the diffracted field using the barycentric method, $r_\Gamma = 2\lambda$.

D_{rc}/λ	$\max_\theta(\text{er})$
5	1.045
10	0.536
20	0.278
30	0.181
40	0.136
50	0.109
60	0.084
65	0.084

Table 2: Evolution of the relative error with respect to the distance to Γ, $r_\Gamma = 2\lambda$.

a larger diffracting obstacle of size $r_\Gamma = 2\lambda$, in the same experimental setting, the barycentric method fails at giving such a good accuracy. Figure 9 displays the map of relative errors of the diffracted field computed by the barycentric method with respect to the reference solution and Table 2 provides values of the relative error.

5.2. The polycentric solver

In this section, we illustrate how the polycentric version of the hybrid strategy improves the accuracy of the computation of the diffraction effects by a disk of diameter $r_\Gamma = 2\lambda$. As expected, the accuracy of the evaluation of the diffracted field on the annulus of receivers is rapidly increased for a sufficient number of partitions. Figure 10 shows the map of relative error for the diffracted field computed using a covering of Γ by 16 subsets. It shows a maximum relative error at 5λ of 2.29%. This covering allows to reach maximum relative error of the same order as that of the ray tracing method. Besides, to show the evolution of the maximum of the relative error with respect to the number of subsets, we conducted several numerical experiments for subdivions of Γ involving 3 to 40 subsets. The evolution of the maximum of the relative error at 5λ from y_c with respect to the number of subsets is plotted on figure 11. We notice that the maximum relative error can be decreased to an asymptotic value of 1%.

6. Conclusion

A hybrid strategy combining the ray tracing method and the integral equation method was here proposed to allow a good high frequency approximation of the diffraction phenomena both in the far field and near field zones. The strategy principle was described on a simplified configuration, where the diffracting obstacle is supposed to be far from the boundaries of an homogeneous propagating medium. Nevertheless, the strategy is also applicable in the more relevant configuration of propagation in a large homogeneous medium containing multiple inhomogeneities. Besides, current work aims at extending the presented approach to cases where

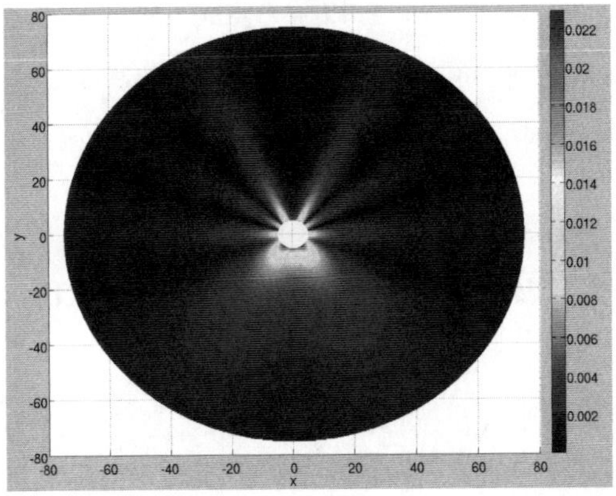

Figure 10: Map of relative error er of the diffracted field using the polycentric method, $r_\Gamma = 2\lambda$, $P = 16$.

D_{rc}	$\max_\theta(\text{er})$ ($\times 10^{-3}$)
5	22.9
10	10.8
20	5.24
30	3.47
40	2.60
50	2,07
60	1.73
65	1.59

Table 3: Evolution of the relative error of the polycentric evaluation of the diffracted field with respect to the distance to Γ, $r_\Gamma = 2\lambda$, $P = 16$.

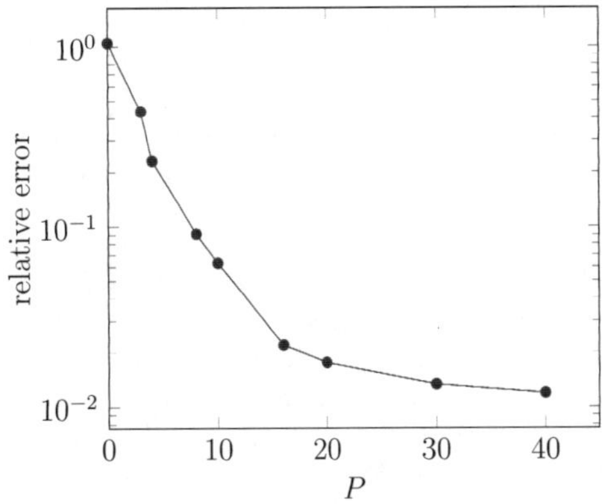

Figure 11: Evolution of the maximum relative error of the diffracted field at 5λ from y_c with respect to the number P of subsets used in the covering of Γ.

the defect is close to a boundary. To conclude, the approach presented is generic and can be extended to 3D elastodynamics, which is our ultimate objective.

References

[1] Watson N G 1966 *A treatize on the theory of Bessel functions* (Cambridge University Press) p194-224

[2] Nedelec J C 2001 *Acoustic and Electromagnetic Equations. Integral representations for harmonic problems* (Applied Mathematical science vol 144) ed Marsden J E and Sirovich L (New-York: Springer-Verlag)

[3] Colton D and Kress R 1998 *Inverse Acoustic Electromagnetic Scattering Theory* (Applied Mathematical Sciences **98**) (Heidelberg: Springer) chapter 3 p37-84

[4] Keller J B and Lewis R M 1995 Asymptotic methods for partial differential equations: the reduced wave equation and the Maxwell's equations, *Surveys in applied mathematics* (Springer US) chapter 1 p1-82

[5] Cerveny G 2000 *Seismic ray theory* (Cambridge University Press)

[6] Runborg O 2007 Mathematical Models and Numerical Methods for High Frequency Waves *Comm. in Comp. Physics* **2** p 827-880

[7] Barbone P R and Michael O 1999 Scattering from submerged objects by a hybrid asymptotic-boundary integral equation method *Wave motion* **29** p137-56

[8] Rokhlin V and Yarvin N 1998 Generalized Gaussian Quadratures and Singular Value Decompositions of Integral Operators *J. on sc. Comp. and Appl. Math.* **20** p699-718

IOP Conf. Series: Journal of Physics: Conf. Series **1017** (2018) 012008 doi:10.1088/1742-6596/1017/1/012008

Simulation of the spatial frequency-dependent sensitivities of Acoustic Emission sensors

N Boulay[1], A Lhémery[1] and F Zhang[2]

[1] CEA LIST, F-91191, Gif-sur-Yvette, France
[2] CETIM, F-60304, Senlis, France

E-mail: alain.lhemery@cea.fr

Abstract. Typical configurations of nondestructive testing by Acoustic Emission (NDT/AE) make use of multiple sensors positioned on the tested structure for detecting evolving flaws and possibly locating them by triangulation. Sensors positions must be optimized for ensuring global coverage sensitivity to AE events and minimizing their number. A simulator of NDT/AE is under development to provide help with designing testing configurations and with interpreting measurements. A global model performs sub-models simulating the various phenomena taking place at different spatial and temporal scales (crack growth, AE source and radiation, wave propagation in the structure, reception by sensors). In this context, accurate modelling of sensors behaviour must be developed. These sensors generally consist of a cylindrical piezoelectric element of radius approximately equal to its thickness, without damping and bonded to its case. Sensors themselves are bonded to the structure being tested. Here, a multiphysics finite element simulation tool is used to study the complex behaviour of AE sensor. The simulated behaviour is shown to accurately reproduce the high-amplitude measured contributions used in the AE practice.

1. Introduction

Acoustic Emission (AE) is a widely used nondestructive testing (NDT) technique, known to be very sensitive to early damaging processes in possibly very large industrial structures. It is a passive technique where the source of elastic waves is typically an evolving defect and where AE sensors are positioned (possibly permanently attached) onto the structure. The number and the positions of sensors must be optimized so that any event can be detected wherever it happens in the component under test (full coverage must be ensured) with a minimal number of sensors for obvious economic reasons (number of independent channels to manage etc.). The method is known to offer high detection capabilities. It can further be used for defect location. For this, several sensors must detect the same AE event, and measure the arrival time at each of these sensor positions. Then, assuming the wave speed in the structure is known, triangulation algorithms can be applied to locate the source of the AE event.

However, typically measured AE time-dependent signals are complex. Physical phenomena taking place are of various nature and involve various space and time scales, from [µm] and less than [µs] for crack propagation to [m] and more than [ms] for wave propagation. Accounting for a single wave speed in the structure – a necessary assumption for applying simple triangulation – often constitutes an over-simplification of actual elastic wave propagation in an industrial structure, all the more since waves propagating as guided waves are involved (in thin structures). For these reasons, AE testing is not considered as being a quantitative nondestructive technique. Despite the non-quantitative nature of the technique, the availability of a simulation tool that can address this complexity would constitute a useful mean for optimizing AE testing configurations and for helping the interpretation of measured signals. The development of such tools attracts far less people than that of tools for simulating more quantitative techniques (ultrasonic, eddy-current, radiography). Several objectives of such developments can be identified. A first objective is to get accurate knowledge about influencing parameters among the many parameters involved [1]; simplifications can be considered so that some parameters can be treated independently of the others. Once this objective is reached, tools able to deal with realistic configurations must be developed to attain the second and more ambitious objective of predicting typical signals at several sensor locations, given one AE event, one component under test and assuming sensor sensitivity is known. This implies to model AE events (for example crack evolution in metallic parts [2], damage in composites [3]), elastic wave propagation in a component [4] and the behaviour of AE sensors [5]. References 1, 2, 4 and 5 correspond to a first attempt at CEA to pave the path for the development of a simulation tool of AE testing to be included in the CIVA platform [6].

The present paper addresses the problem of predicting sensor sensitivity. Evidences of a complex behaviour of standard AE sensors led to the development of various methods to characterize them experimentally over the last decades. A recent review can be found [7] where results obtained by the main methods and applied to the same sensors are compared. Clearly, each method accesses to some characteristics of the sensor behaviour, but not all to the same ones. In other words, the characterization obtained by these various methods is incomplete. In the present study, our goal is to predict this behaviour (as done in the case of conical AE sensor [8]) to understand what makes it complex, to check experimentally the validity of predicted results to further deduce from this knowledge a strategy to include sensor simulation in a global AE simulation tool.

2. Sensitivity of acoustic emission sensor

To predict the sensitivity behaviour of an AE sensor, we first briefly discuss the notion of sensitivity and assumptions or even simplifications made in the AE community about it.

A typical AE sensor can be regarded as a linear system with a mechanical input (particle velocity, particle displacement, force…), and with a voltage as output. Its sensitivity function is used to describe its voltage response to an arbitrary incoming wave. Sensor sensitivity is said to be measurable by means of calibration methods such as step-force calibration, calibration by laser interferometer, face-to-face calibration or reciprocity calibration [7].

Sensors are supposed to respond to a weighted average of the mechanical input, relating to diffraction effects due to its finite aperture [9]. The sensitivity is described as a spatially dependent function over the sensor front face S in contact with the part. A classical form to account for aperture effects writes

$$V(t) = \frac{1}{S}\iint_S \mathbf{A}(\mathbf{r}) \cdot \mathbf{u}(\mathbf{r},t) dS\,, \tag{1}$$

where $\mathbf{r}$ denotes a running position at S, t the time, $V(t)$ the open circuit voltage, $\boldsymbol{A}(\mathbf{r})$ the local sensitivity vector and $\mathbf{u}(\mathbf{r},t)$ the particle displacement associated to the incoming wave at the surface. Further, in most cases, only the normal component of the displacement is considered. This writing means that $\boldsymbol{A}(\mathbf{r})$ is not considered to be time (or frequency) dependent. A common approximation considers that the sensitivity is scalar and uniform [10], $\boldsymbol{A}(\mathbf{r}) \sim A_0$. Under these approximations, Eq. (1) reduces to

$$V(t) \approx \frac{A_0}{S}\iint_S u_z(\mathbf{r},t) dS\,. \tag{2}$$

This rather rough simplification for local sensitivity function is generally justified by the fact that, since most AE sensors are axisymmetric, tangential components of a normally incident wave vanishes by averaging effect so that only its normal component makes an effective contribution. Deviations between experimental and computed aperture function in [10] probably comes from the strong assumption of uniform local sensitivity. A less restrictive assumption has been used in [5] by considering the local sensitivity as a function of the position at the sensor surface leading to improve the theoretical expression of aperture function. In that case, local sensitivity can be written as

$$A(\mathbf{r}) = A_0\psi(\mathbf{r})e^{-i\varphi(\mathbf{r})}, \tag{3}$$

where A_0 is the maximum sensitivity to normal displacement of the sensor surface, $\varphi(\mathbf{r})$ the phase shift of this displacement and $\psi(\mathbf{r})$, a weighting function. The output voltage becomes:

$$V(t) \approx \frac{A_0}{S}\iint_S \psi(\mathbf{r})e^{-i\varphi(\mathbf{r})}u_z(\mathbf{r},t)dS. \tag{4}$$

In the next section, we will show that previous approximations are not valid. As so, calibration methods designed to characterize over-approximated sensitivities lead to experimental uncertainties.

3. Finite element modelling of AE sensor

3.1. Finite element configuration

To understand the actual behaviour of AE sensors, COMSOL Multiphysics finite element (FE) software has been used to carry out FE simulations [11]. The sensor being studied is a typical resonant one, developed by CETIM, and shown in Fig. 1. A thick piezoelectric element (PZT) is glued inside the sensor case onto which a cap is screwed. In the configuration of computation, the sensor is coupled to a hemispherical elastic medium surrounded by a perfectly matched layer (PML) to block wave reflections; this models a semi-infinite elastic solid. Classical PML for elastodynamics as implemented in COMSOL are used for computations in the frequency domain. Their validity has been very simply checked by means of numerical experiments showing that an outgoing wavefield does not give rise to reflected contributions from the interface between the modelled elastic medium and the PML.

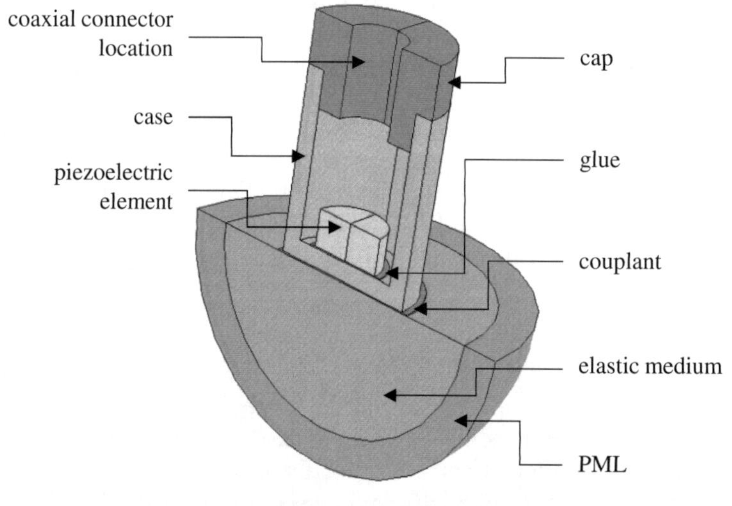

Figure 1. Configuration considered for FE computation of AE sensor behaviour

AE sensors without preamplifier are reversible and can be used as a transmitter. Calibration methods based on reciprocity make use of this possibility. It is also useful for studying the sensor behaviour by the FE method. Generally speaking, FE studies where elastic wave phenomena are involved are costly in computation time and in memory because of the need for finely meshing the whole configuration, notably to account for wave propagation in thin layers. Considering a part of large volume comprising an AE source distant from the sensor to be modelled with a full account of 3D effects would be difficult to deal with. Conversely, modelling the sensor behaving as an emitter allows us to reduce to a minimum the meshed elastic medium to which it is attached. Indeed, the active source is the sensor itself so that a small volume for modelling the part is enough to take into account its mechanical influence on sensor behaviour, once non-reflecting boundary conditions are chosen to close the volume at a short distance. Furthermore, considering the behaviour of the sensor acting as an emitter allows us to account for its axisymmetry, thus, to considerably reduce the size of the FE system to solve and the associated computation time, by modelling the whole configuration by means of axially symmetric elements.

In such a configuration, the various modes (thickness-modes, radial modes and more complex modes involving radial and thickness coupling) of the thick piezo-element are excited. In what follows, to finely capture the complex behaviour of the sensor, simulations are mostly carried out in the frequency domain over the typical bandwidth of AE sensor (1-1000 kHz), though computations in the time-domain are also made when full-waveforms are predicted and compared to experimentally measured signals.

3.2. Uniform sensitivity and contactless sensitivity measurement
In [5], non-uniformity of the sensitivity over the sensor front face has been demonstrated experimentally. In these measurements, an AE sensor is used as a transmitter radiating in air (unloaded); contactless point measurements of time-dependent velocity over the sensor surface are made by laser velocimetry. Then, a Fourier transform is performed to analyse it over the bandwidth of interest.

The FE model allows us to simulate such an experiment directly in the frequency domain. Simulation is also used to compute the velocity at the same points when the sensor is bonded to a solid (loaded case). Figure 2 compares the modulus of the particle velocity at 419 kHz for the unloaded case and for the loaded one. Obviously, the local velocity is far from being uniform as already mentioned in [5]. Now, it also clearly appears that the unloaded and the loaded cases lead to two different profiles.

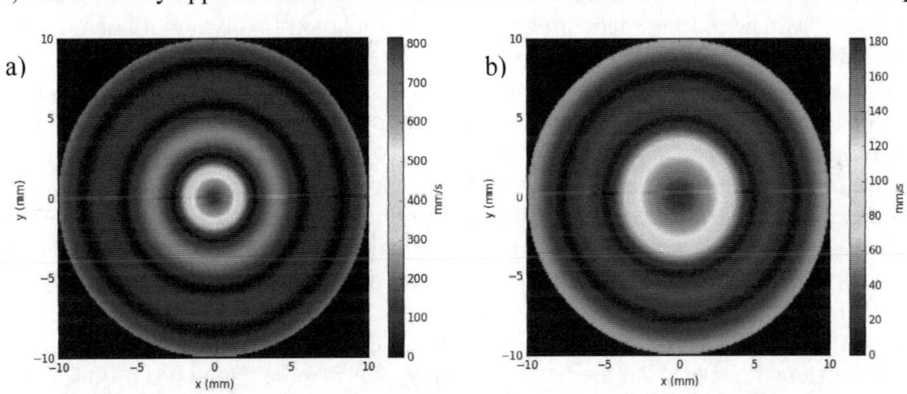

Figure 2. Local velocity modulus at the sensor surface for an excitation frequency of 419 kHz. a) the sensor is free, b) the sensor is in contact with semi-infinite medium made of steel

Another way to illustrate the difference between these two cases is shown by Fig. 3 which represents the normal particle velocity averaged over the sensor face as a function of the excitation frequency.

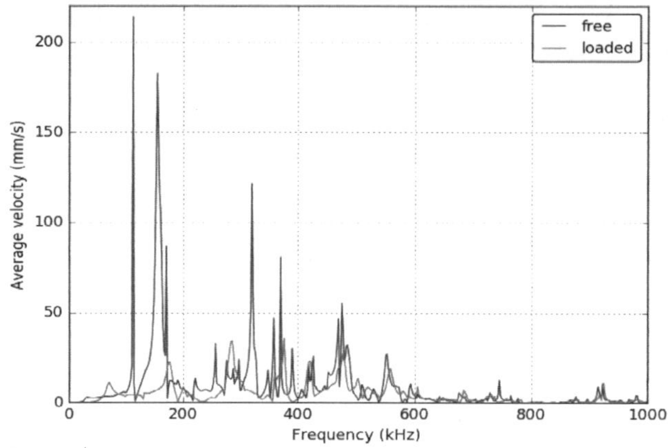

Figure 3. Averaged velocity over the sensor face as function of the frequency. The sensor is either mechanically loaded (red) or free (blue)

The loaded case leads to a response which is smoother and of lower amplitude than the free case. In the loaded case, transmission from the sensor to the elastic medium is high (small impedance mismatch). In the free case, waves in the sensor are trapped since they cannot be radiated due to the huge impedance mismatch, this leading to both higher amplitude of waves and further resonances.

3.3. Influence of radial displacement

Considering results obtained for the loaded case can improve accuracy in determining the actual aperture function of the sensitivity as defined in [5], though a method for its measurement is to be invented for accessing to local mechanical quantities in the presence of a solid part. However, one of the main assumptions made in the literature and in the formulation involving an aperture function is that the sensor is sensitive only to normal component.

Now, AE sensors comprise cylindrical piezo-elements of radius approximately equal to their thickness, glued to the sensor case. Resonances of such elements are not restricted to a harmonic series of thickness resonances but inevitably include others involving radial modes or coupling radial and thickness vibration modes. Sensors are therefore sensitive to both normal and tangential contributions associated to incoming waves. In order to compare the influence these modes of the piezo-element on sensor behaviour, one can compute the ratio between the averaged radial and the averaged normal displacements along a radius of the sensor surface over the whole bandwidth, generated by the sensor used as an emitter. If the average was computed over the surface rather that along the radius, the averaged tangential displacement would be equal to zero as the sensor is axisymmetrical.

Two cases are shown in Fig. 4. In the first case, the sensor is mounted on a semi-infinite medium, whereas in the second one, it is mounted on an infinite 3-mm-thick plate. In both cases, the frequency-dependent above-defined ratio is shown together with a "smoothed" version of it to ease its reading. The smoothing is obtained by applying a classical Savitzky-Golay filter, as done in [7].

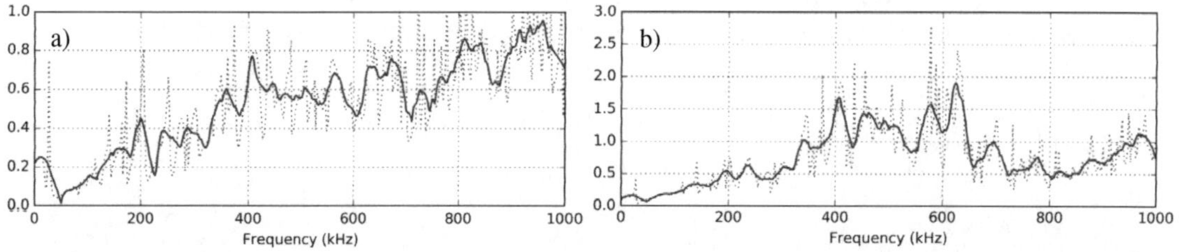

Figure 4. Ratio (dotted line) and "smoothed" ratio (thick line) of averaged radial over averaged normal displacement as a function of the frequency. a) Sensor mounted on a semi-infinite medium, b) sensor mounted on an infinite plate.

At first, one can observe that the averaged radial displacement is of the same order of magnitude as the normal one. Interestingly, depending on the geometry of the medium in which the transducer radiates, frequency-dependent ratios differ. In the plate, the radial displacement is even higher than the normal displacement in the range 400 to 600 kHz. Thus, it can be more accurate to consider not only the normal displacement in calibration techniques but also the radial one.

Finally, the most complete sensor response that can be simulated accounts for both normal and tangential sensitivities, for their nonuniform value along the radius of the sensor front face, and for their variation with the frequency. These sensitivities are obtained by computing in the frequency domain the surface traction $\mathbf{T}_E(\mathbf{r}, f)$ at the interface between the AE sensor and the part (modelled as a semi-infinite medium), by considering that the sensor is used as an emitter, excited by an alternative current I_E (same amplitude over the bandwidth [0;1] MHz). We have shown [12], using the electromechanical reciprocity principle [13], that the output voltage produced by an incoming wave of particle velocity $\mathbf{v}_R(\mathbf{r}, f)$ incident on the sensor surface, can be written as

$$V(f) = \frac{1}{I_E} \iint_S \mathbf{T}_E(\mathbf{r}, f) \cdot \mathbf{v}_R(\mathbf{r}, f) dS. \quad (5)$$

According to Eq. (1), the sensitivity function $\mathbf{A}(\mathbf{r}, f)$ written in the frequency domain is given by

$$\mathbf{A}(\mathbf{r}, f) = \frac{i2\pi f S}{I_E} \mathbf{T}_E(\mathbf{r}, f). \quad (6)$$

It is not the purpose to give the full demonstration of this theoretical result which is out of the scope of the present paper. The two sensitivity functions (normal and tangential) are shown on Fig. 5 as two colour maps of their amplitude (in dB) as functions of the frequency and of the distance of the running position at the sensor front face from the centre of symmetry. The tangential sensitivity is displayed relatively to the maximum of the normal sensitivity to ease the reading of their relative values.

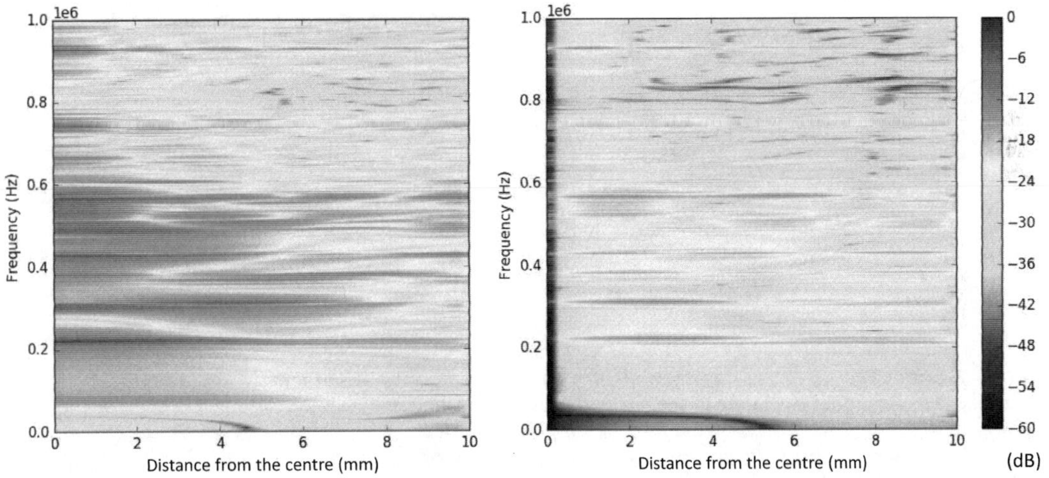

Figure 5. Normal (left) and radial (right) sensitivities (in dB) given as functions of the position at the sensor surface (distance from the centre in mm) and of the frequency (over the bandwidth [0; 1] MHz).

These results demonstrate once more that the sensitivity functions are not spatially uniform, are frequency-dependent and that the sensitivity to tangential displacement cannot be neglected.

4. Experimental validation

In order to perform some experimental validations of our FE computations, we set-up the experiment schematized by Fig. 6. An AE sensor used as a transmitter is mounted on a 3-mm-thick aluminium plate. Normal displacement at the surface of the plate is measured by a laser interferometer which

IOP Conf. Series: Journal of Physics: Conf. Series **1017** (2018) 012008 doi:10.1088/1742-6596/1017/1/012008

scans the plate surface at a set of points of a Cartesian 2D grid. At each position, the absolute time-dependent waveform measured by the interferometer is recorded.

Radiation into a plate is a more complex case than that into a massive elastic solid as, over the frequency bandwidth considered, elastic waves radiated in the plate propagate as guided waves which are multimodal and dispersive. This configuration is however of interest as many practical cases relevant to AE testing concern "thin" structures which involve propagation of elastic guided waves.

A snapshot of the measured normal displacement is displayed in Fig. 6. The emitter is at the centre of the map. The presence of the coaxial cable used for electrical excitation to the sensor manifests as a dark vertical line between the sensor and the bottom of the map. Figure 6 also displays the electrical excitation used in both the experiments and for the simulated results which follow.

Let us now compare simulated and measured waveforms. Since the area of interest is around the sensor position, Low-Reflecting Boundaries (LRB), specific to time-dependent computations in COMSOL [11], are introduced to absorb outgoing waves so that no edge-reflected contribution can superimpose with waves directly radiated by the source. In the experiment, the plate is large but of course finite so that waves reflected at the plate edges eventually superimpose with the directly radiated field. Therefore, the signal recorded in the experiments is time-windowed so that only the direct contributions are recorded.

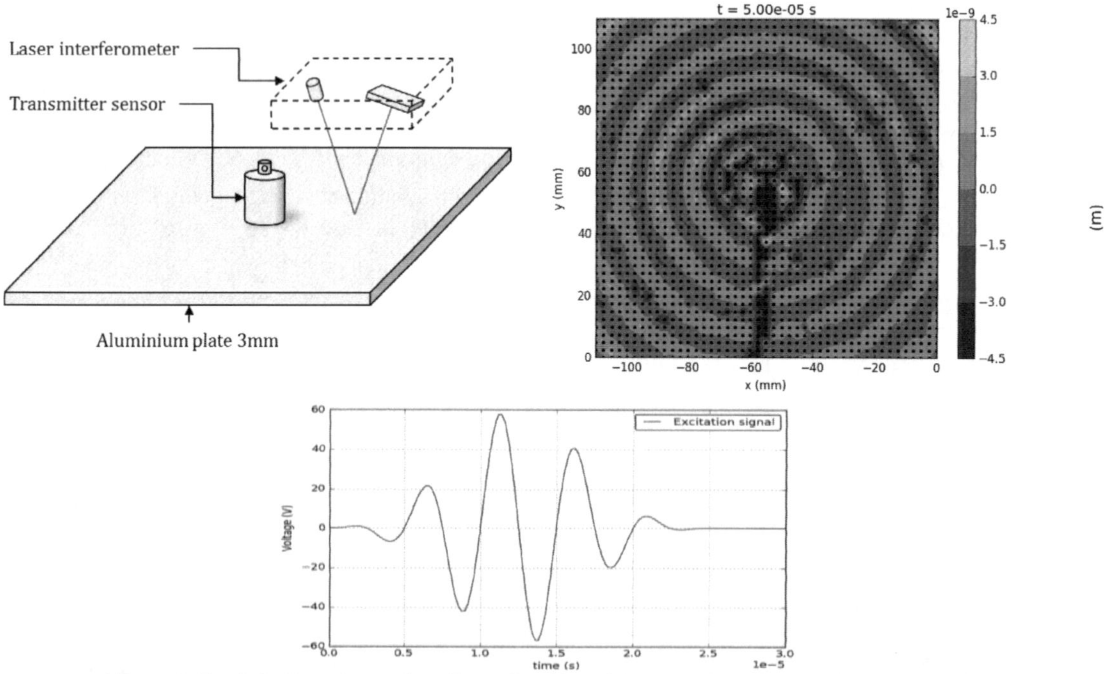

Figure 6. Top-left: Experimental configuration. Top-right: snapshot of normal displacement. Bottom: time-dependent excitation of the AE sensor used as an emitter (in simulations and in experiments).

Figure 7 compares simulated and measured waveforms for the normal particle displacement of the top surface of the plate at a point 50 mm distant from the source centre. The two waveforms are displayed in absolute amplitude, the same amplitude of electrical excitation being considered for both results.

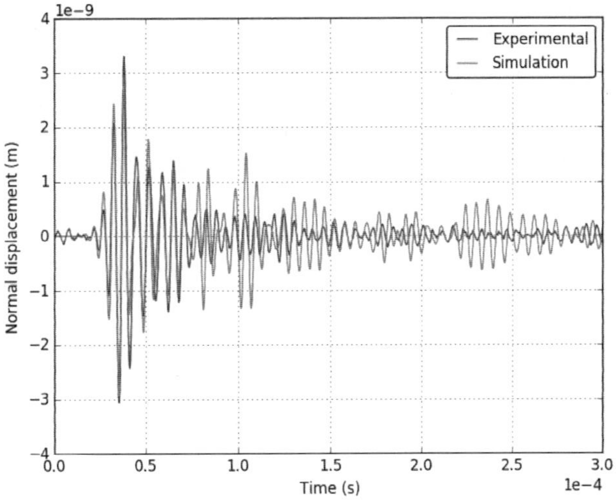

Figure 7. Simulated (red) and measured (blue) waveforms of normal particle velocity at a point distant of 50 mm from the centre of the AE sensor

The first wave packets are very well predicted by the simulation, both qualitatively for the shape of the waveform and quantitatively for its absolute amplitude. This clearly demonstrates the overall relevance of the ways the sensor, its excitation and the wave radiation are modelled. The accuracy of prediction for the wave packets arriving at longer time-of-flight is however far lower.

It is quite difficult to identify the very reasons for this lack of accuracy in the later packets. There are uncertainties in the values of some parameters. A sensitivity analysis was carried out where the values of selected model input parameters were varied. For example, the influence of glue thickness inside the sensor case or that of the elastic characteristics of the thin coupling medium between the sensor and the part have been studied. None of them had a quantitatively significant impact and the details of this sensitivity analysis are beyond the scope of this paper.

The actual sensor geometry is not exactly axisymmetric (flat surfaces are machined to hold the sensor more easily), as shown in Fig. 8, whereas in the computations, as mentioned above, the choice of purely axisymmetric configuration was made as it leads to solve far smaller FE systems. However, the influence of this simplification has been studied: it is very small and cannot explain the lower accuracy of predicted secondary wave packets.

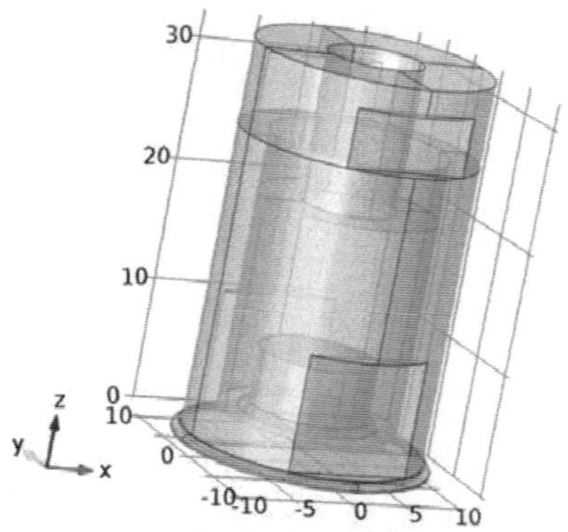

Figure 8. Actual geometry of the AE sensor used in experiments

Other simplifications were made to model the overall sensor geometry in the initial result shown in Fig. 7 where the coaxial cable, screwed to the sensor cap (thus “filling” the hole at its top), was not taken into account. Generally speaking, some elements of geometry may influence significantly the simulation results. To show this, it is interesting to first make an extreme simplification to model the sensor case consisting in not considering it at all. The simulated waveform for the case of a piezo-element directly coupled to the plate is compared to the measured waveform, as displayed in Fig. 9. Interestingly, the order of magnitude is still the same, and the very first oscillations are quite similar. But the predicted waveform shows large resonances that are not present in the measured signal, thus demonstrating that the sensor case has an obvious role in the process.

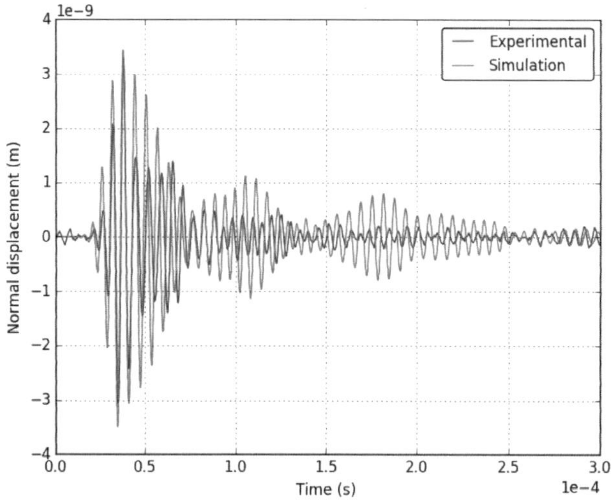

Figure 9. Simulated (red) and measured (blue) waveforms of normal particle velocity at a point distant of 50 mm from the centre of the AE sensor. In the simulation, the piezo element is directly coupled to the plate.

To show this very clearly, a map of the particle displacement is given on Fig. 10, for a CW excitation at the frequency of 150 kHz. Again, since we are now in the frequency domain, PML boundary conditions are used at the top of the sensor connector and cable as well as in the plate. The particle displacement has significant amplitude in the PZT element but also in the sensor case.

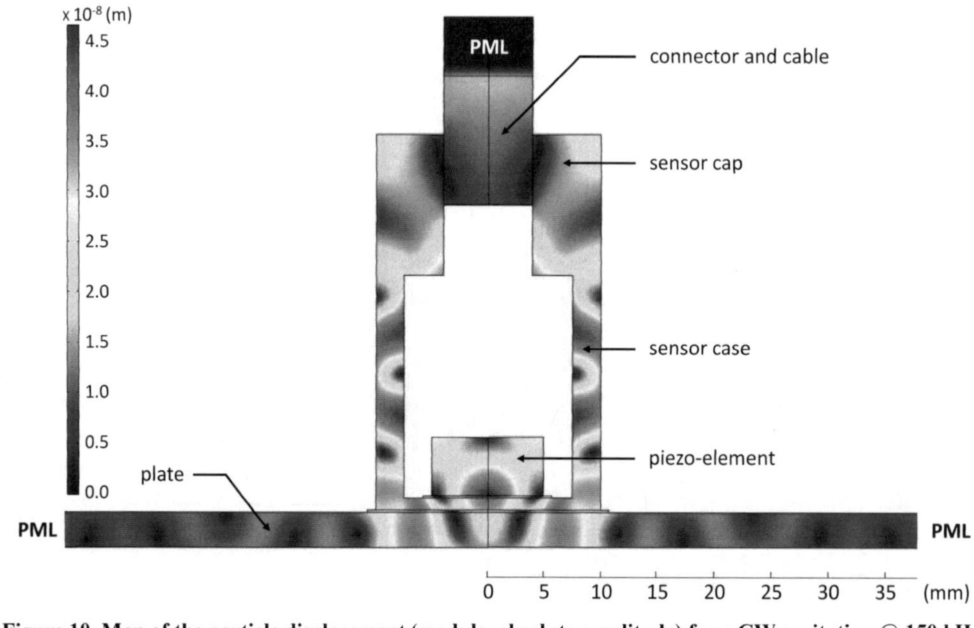

Figure 10. Map of the particle displacement (module, absolute amplitude) for a CW excitation @ 150 kHz.

Now, if the cap modelled for the result given by Fig. 7 is replaced by artificial boundary conditions (LRB in the time domain), the predicted waveform shown in Fig. 11 differs from that accounting for the cap (but not accounting for the coaxial cable). Some energy is lost in the LRB layer which otherwise would propagate through the cap to eventually contribute to the radiated field.

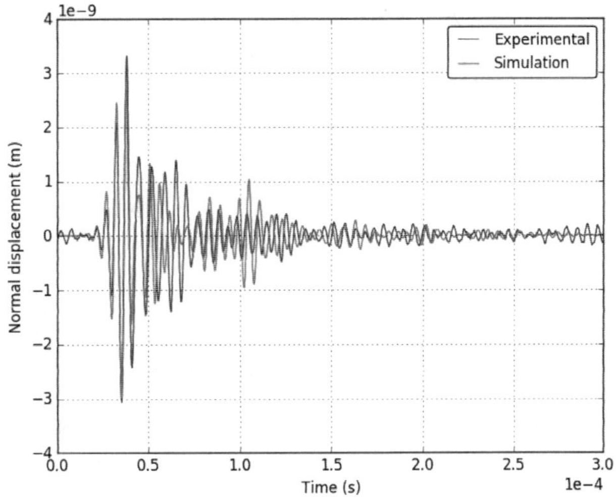

Figure 11. Simulated (red) and measured (blue) waveforms of normal particle velocity at a point distant of 50 mm from the centre of the AE sensor. In the simulation, the sensor cap is replaced by PML boundary conditions.

This time, there are discrepancies at the end of the first packet (between 50 and 75 μs) but the amplitude of secondary packets (after 100 μs) agrees better with that of the measured waveform.

These simulation results establish that the field radiated by the sensor, and by reciprocity, the sensitivity of the sensor, strongly depends on the geometry of the sensor case. The geometry must therefore be accounted for, if an accurate prediction of its behaviour is sought. In the present case, treated as an example to study the complex behaviour of AE sensors, the actual behaviour would probably be better predicted by accounting for the complex structure of the cap, by including the presence of the coaxial cable and the way it is screwed to the case, the cap itself being screwed to the sensor case.

5. Discussions and conclusions

The simulation study by means of FE computations highlighted several salient characteristics of the behaviour of typical AE sensors.

First, their simple design leads to complex behaviour. The broadband sensitivity of AE sensors is obtained through the use of a thick cylindrical piezo-element, a geometry leading to a large number of eigen-modes in the frequency band of interest. Resonances are not smoothed by backing the element, a standard way for broadening the bandwidth of an ultrasonic transducer with a thin piezo-element typical of those used in ultrasonic testing; the smoothing of resonances results from complex propagation into the sensor case in which the element is glued. As waves propagate into the sensor case, the case design itself affects the overall sensitivity through a complex coupling of the piezo-element with the other elements constituting the sensor.

As a result of the above factors, the sensitivity is not simple to describe. Idealized AE sensor behaviour modelled as a uniform sensitivity to normal displacement, possibly refined by taking into account a variation with frequency, is too strong an approximation of the actual behaviour of AE sensor. Rather, it is nonuniform over the sensor front face in contact with the part under examination; furthermore, the nonuniform function of position is frequency-dependent. Furthermore, it depends on the mechanical loading exerted by the part on which the sensor is attached.

Finally, in order for it to be thoroughly described, the sensitivity function must be defined as a vector function comprising tangential and normal components which are complex-valued, which depend on two variables, namely, the frequency and the position over the surface in mechanical contact with the part. As the mechanical loading modifies the sensitivity, a better image would be that of a family of functions, parametrized by mechanical characteristics of the piece under test.

Standard experimental methods of AE sensor characterization only partially capture the complex behaviour of these sensors. Knowledge gained in accurately predicting this behaviour will help devise new methods for a full characterization of AE sensors.

References

[1] Lhémery A, Ben Tahar M, Foucher F, Mesinele A, Recolin P, Zhang F 2011 COFREND Conference Proceedings http://www.ndt.net/article/cofrend2011/papers/237.pdf,
[2] Hello G, Ben Tahar M, Roelandt J-M 2012 *Int. J. Solids Struct.* **49** 556
[3] Sause M G R 2016 *In Situ Monitoring of Fiber-Reinforced Composites - Theory, Basic Concepts, Methods, and Applications* (Springer Int. Publishing, Switzerland)
[4] Ben Khalifa W, Jezzine K, Grondel S, Hello G, Lhémery A 2012 *J. Acoust. Emission* **30** 137
[5] Monnier T, Dia S, Godin N, Zhang F 2012 *J. Acoust. Emission* **30** 152
[6] Details about capabilities and uses of CIVA platform: http://www.extende.com/
[7] Ono K *Materials* 2016, **9**, 508 (corrections: Ono K *Materials* 2016, **9**, 784)
[8] Sause M G R, Hamstad M A, Horn S 2012 *Sensors and Actuators A: Physical* **184** 64
[9] ASTM 1987 Non Destructive Testing Handbook: Acoustic Emission Testing 2nd edition (Columbus, OH: American Society for Non Destructive Testing) pp 126-130
[10] Goujon L and Baboux J-C 2003 *Meas. Sci. Technol.* **14** 903
[11] COMSOL Multiphysics ® 2015. Version 5.2. COMSOL, Inc., Burlington, MA, USA. Users' guide
[12] Boulay N, Lhémery A 2017 in preparation for *NDT&E Int.*
[13] Auld B A 1979 *Wave Motion* **1** 3